THE GREAT
BRAIN DEBATE

THE GREAT BRAIN DEBATE

NATURE OR NURTURE?

John E. Dowling

Princeton University Press
Princeton and Oxford

Published by Princeton University Press, 41 William Street, Princeton, New Jersey
08540
In the United Kingdom: Princeton University Press, 3 Market Place, Woodstock,
Oxfordshire OX20 1SY

Originally published by Joseph Henry Press in 2004
First Princeton paperback edition, 2007

Library of Congress Control Number 2007926585

ISBN: 978-0-691-13310-2

British Library Cataloging-in-Publication Data is available

This book has been composed in TrumpMediaeval-Roman

Printed on acid-free paper. ∞

press.princeton.edu

Printed in the United States of America

1 3 5 7 9 10 8 6 4 2

February 7, 2008

For the next generation—
Madison, Quincy, Grace, and Olivia

CONTENTS

INTRODUCTION

The United States Congress designated the 1990s as the Decade of the Brain, but some suggest that the twenty-first century will be the century of the brain, when the last great frontier in biology —an understanding of the most complex biological system, the human brain—will be breached. Already the considerable advances made in neuroscience over the past 50-100 years are being called upon to explain many things about human behavior. Interdisciplinary programs are appearing in our colleges and universities asking what various disciplines and fields can learn from neuroscience and vice versa. At Harvard, I have been associated with the Mind, Brain and Behavior program since its inception in 1993, and I codirected it for a year. It attracts faculty from the Harvard Medical, Law, Divinity, and Business Schools as well as the School of Education and the Faculty of Arts and Sciences. Fields as diverse as philosophy, music, English, linguistics, anthropology, and history of science are represented, as well as the expected fields of biology, psychology, and computer science.

Many examples can be offered to illustrate the impact of neuroscience on other disciplines; I offer two here. First, studies of how we learn and remember things have demonstrated convincingly that memories are largely reconstructive and creative. False memories are not uncommon. These findings have fundamentally changed the way the law views eyewitness testimony. Contrary to the long-held belief that an eyewitness can faithfully record and remember an event, we now realize that what we remember or even perceive of an event depends on many factors—previous experiences, biases, attention, imagination, and so forth. Different eyewitnesses can give very different reports, though in each case describing what each observer firmly believes he or she saw.

A second example is the placebo effect—long thought to be without physiological basis. If a sugar pill is administered to someone experiencing pain, that person reports a lessening of the pain if told the placebo will help. We now know that the pain reduction is caused by the release of endogenous opiate-like substances in the brain. No drug trial today is carried out without a control cohort receiving a similar, but presumably inactive, agent. But placebo effects can greatly influence the outcome of such trials. How then do we decide what is efficacious and what is not? This question has enormous implications for medical therapies.

How far does the influence of neuroscience extend? Have studies on the developing brain, for example, told us much about how we should raise or educate our children? Some say yes, but others respond with a resounding no. The stakes are high—public programs such as Head Start, costing millions, if not billions, of dollars, are linked to notions supposedly neurobiologically based, but often the neurobiological evidence cited in support of one position or another is weak, controversial, or overinterpreted. The view that the young brain is more modifiable than the adult brain—which is certainly true—led to the notion that the first three years are the essential ones for raising a healthy, happy, and competent child. This extreme view, and the evidence on which it is based, has recently been critically examined in John Bruer's book *The Myth of the First Three Years*. As Bruer clearly documents, the first three years are important for brain development, but so are subsequent years. Nothing closes down completely after just three years—indeed, the brain continues to mature until the ages of 18-20, as we shall see.

What about the adult brain? How hard-wired is it? Once it is injured, is recovery possible or are we stuck with just what was there before the injury? Recent studies suggest that the adult brain is much more plastic than was long believed, but how much plasticity can there be? What about the influence of genes on behavior? How do genes and behavior relate? This contentious subject has generated volumes—with highly polarized views. The list of books written about it is long and includes provocative titles such as *The Mismeasure of Man, Not in Our Genes*, and most recently *The Blank Slate*.

And finally, the aging brain. Does the brain eventually fail in all of us, or is this a pessimistic view? Is it likely that maximal life span can be extended to 150-180 years? What about the age-related neurodegenerative diseases such as Alzheimer's and Parkinson's diseases? Are there reasonable approaches that might be taken to deal with these frightening and devastating conditions?

The purpose of this book is to lay out many of the neurobiological facts we have about the developing, adult, and aging brain. Clearly, the neurobiology is at a primitive stage compared to the richness of psychological observations that have been made on children, adults, and aging people. Nevertheless, not only have modern neurobiological studies given us some firm facts with which to ponder many of the issues laid out above, but neuroscience studies have also given us models—ways to think neurobiologically about the issues. The models in their details might not turn out to be right, but they suggest that we can get at many of the underlying phenomena and understand them.

Ultimately, we seek to understand the human brain, but our ability to study it neurobiologically is limited for the most part to noninvasive imaging or recording techniques. Occasionally we can get a piece of human brain to analyze, but this is the exception. On the other hand, we can study the brains of animals, and often the animal brain data are directly relevant to an aspect of human brain function or, at the very least, they give us a way to think about how the human brain might work. Throughout this book, I give examples of animal brain studies and what I believe they are telling us.

The book is not written for the expert, but for those non-experts and nonscientists interested in the issues and how they are being approached. I have tried to portray the neurobiology fairly and accurately, but in a simplified way. The book is divided into three parts: I, The Developing Brain; II, The Adult Brain; and III, The Aging Brain. Three chapters comprise the section on the developing brain, two the section on the adult brain, and just one on the aging brain. This division reflects to a considerable degree the amount of research and focus on these three aspects of human brain biology. The emphasis might be shifting somewhat as our population ages and the devastation of the age-related neurodegenerative diseases looms greater. Nevertheless, the challenge of understanding how the brain develops and how that understanding might help in raising the next generations to the best of our and their abilities is key to the future of humankind.

Initial work on the book took place during a delightful stay at the Rockefeller Study and Conference Center in Bellagio, Italy. Much of the book was written during an equally delightful stay at the International Institute for Advanced Study in Kyoto, Japan. Lisa Haber-Thomson and Carla Blackmar expertly drew the figures, and Stephanie Levinson provided the crucial secretarial help needed to bring the project to fruition. Jerome Kagan, Mark Konishi, Brian Perkins, and Richard Sidman read parts or all of the manuscript and provided many useful corrections, comments, and suggestions. And last but not least, Jeffrey Robbins enthusiastically encouraged the book, edited it, and improved it immeasurably.

PART I

THE DEVELOPING BRAIN

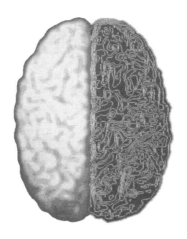

BUILDING A BRAIN

Understanding how the brain forms is one of biology's greatest challenges. From a relatively few undifferentiated cells in the young embryo, all of the neurons and glial (supporting) cells arise. The adult human brain contains about 100 billion neurons (a conservative estimate) and perhaps 10 times as many glial cells. Because virtually all neurons and most glial cells form before we are born, an embryo would generate approximately 250,000 cells per minute in the womb if brain cell generation were constant over the nine-month gestation period. However, most neurons are generated in the first four months of gestation, so the number of cells generated per minute during this early period is much higher. Furthermore, many brain regions initially overproduce neurons and the surplus dies during the maturation process. Thus, at various times during the gestation period more than 500,000 cells might be generated per minute!

Our brain begins to form about three weeks after conception. A group of about 125,000 cells forms a distinctive flat sheet along

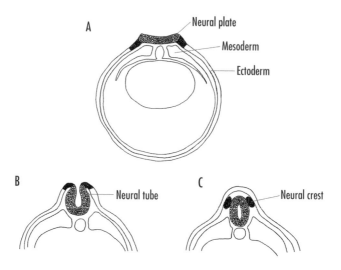

FIGURE 1-1 Formation of the neural plate, neural tube, and neural crest in young embryos.
A: The neural plate cells derive from ectodermal cells on the dorsal surface of the embryo. Signals coming from underlying mesodermal cells induce the dorsal ectodermal cells to become neural plate cells.
B: The neural plate invaginates to become the neural tube, and cells that initially lie laterally along the neural plate form the neural crest.
C: The neural tube becomes the central nervous system (brain and spinal cord), whereas the neural crest forms much of the rest of the nervous system (peripheral nervous system).

the dorsal or back side of the embryo. Known as the neural plate, all the neurons and glial cells derive from this early structure (Figure 1-1A).

Between the third and fourth weeks of development, the neural plate curves inward and creates a groove that slowly closes into a long tube, the neural tube, as shown in Figure 1-1B. The entire central nervous system—that is, the brain and spinal cord—develops from the neural tube. The anterior part of the neural tube becomes the brain proper, the posterior part the spinal cord. By the 40th day of development, three swellings become apparent along the anterior part of the neural tube as shown in Figure 1-2. These eventually form the three major subdivisions of the brain—the forebrain, midbrain, and hindbrain.

During formation of the neural tube, some cells on either side separate to form structures known as the neural crests as shown in Figure 1-1C. Much of the peripheral nervous system—those nerve and glial cells that lie outside the brain and spinal cord—derive from the neural crest cells.

Figure 1-2 shows the development of the human brain from the neural tube. The detailed drawings at the left are enlarged relative to the drawings on the right. By 60 days after conception, the forebrain, midbrain, and hindbrain regions can be readily distinguished. Infolding or wrinkling of the brain's surface—to increase the cortical area—begins at about seven months.

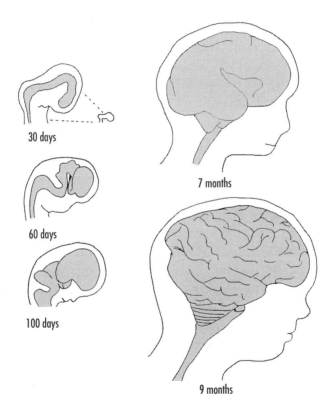

30 days

60 days

100 days

7 months

9 months

FIGURE 1-2 Development of the human brain from the neural tube. The drawings on the left are enlarged relative to those on the right. The tiny drawing at the top indicates the actual size of the brain at 30 days relative to size of the brain at 7 and 9 months.

At nine months of gestation, the brain overall looks quite adult, but it has far to go. The average newborn human brain weighs less than 400 grams, whereas the typical human adult brain weighs about 1,400 grams. Figure 1-3A shows the brain viewed from the dorsal (top) side in a newborn and at six years of age.

Much of the weight increase occurs during the first three years after birth, but the brain does not reach its maximum weight until about 20 years of age. Thereafter, brain weight declines slowly but steadily. Figure 1-3B shows graphically the average weight of the human male brain (based on measurements made on more than 2,000 normal brains) from birth to age 85. Female brains, on average, are slightly smaller at all ages, probably because women tend to be somewhat smaller than men.

I noted above that virtually all neurons are generated by about birth or certainly by six months of age. Thus, what underlies the remarkable growth of the brain in the first three to five years of life? A number of things are going on, including an increase in the number of glial and other supporting cells, growth of blood vessels, and, importantly, the ensheathing of the long axonal processes of the neurons by myelin. Myelin is formed by glial cells wrapping their cell membrane around axons, creating a highly enriched lipid layer that covers the axons. Myelin insulates the axons, making them more efficient in transmitting the electrical signals that travel their length.

However, the most important factor contributing to brain size increase in the early years is the growth and elaboration of the neurons themselves. Not only do their cell bodies grow in size, they also extend more dendritic branches during brain maturation. The dendrites grow larger and go longer distances as shown in Figure 1-4.

More than 80 percent of total dendritic growth probably occurs after birth. It is on the dendrites of a neuron that most synaptic contacts are made; thus, the elaboration of the dendritic processes of neurons that occurs during the brain growth of the

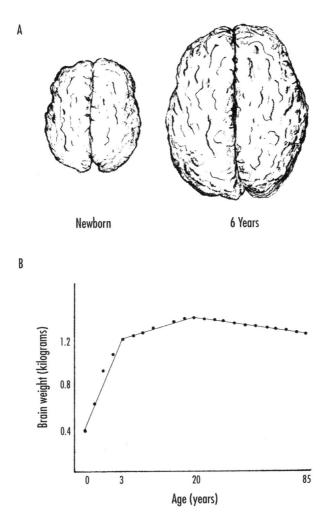

FIGURE 1-3 A: Growth of the human brain from birth (left) to age 6 (right). These are dorsal views showing the cortical surfaces of the brain. B: Brain weight as a function of age. A rapid increase in brain weight occurs in the first three years. The rate of increase then slows, but the brain does not reach its maximum weight until about 20 years. Thereafter, there is a slow and constant loss of brain weight.

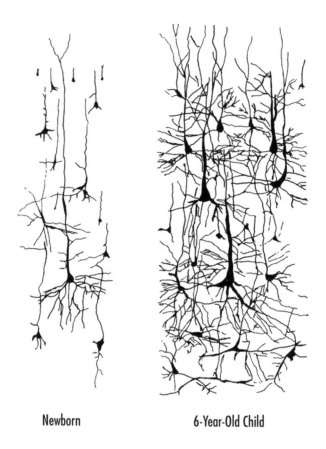

Newborn   6-Year-Old Child

FIGURE 1-4 *The elaboration of neurons during brain maturation. Not only do the cell bodies of the neurons increase in size, but there is an enormous increase in the number, extent, and complexity of their branches.*

early years implies a substantial increase in the synaptic circuitry of the brain.

There is no question that there is an enormous increase in total numbers of brain synapses, not only prenatally but also post-natally up to at least age 2. But the situation is much more complex than just adding synapses. As we shall see, there is a substantial rearrangement and pruning of synapses during brain development and growth, so not only are many synapses added, but many others are lost. Indeed, if one simply looks at the total number of synapses, the peak is between six and eight months postnatally, and then total numbers decline. Experience clearly influences the rewiring of brain synapses during brain maturation, but this rewiring is not just limited to the young brain. All of our lives our brains are being changed by our experiences, and these changes are reflected in the synaptic circuitry of the brain. (See Part II, The Adult Brain, for more information on these changes.) It is certain, however, that the young brain is considerably more plastic than is the adult brain, a topic we shall return to in the next chapter.

It is important also to emphasize that not all parts of the nervous system mature simultaneously. Maturation occurs in a roughly tail-to-head gradient. For example, the spinal cord and brain stem (which controls vital body functions such as respiration, heart rate, and gastrointestinal function) are essentially fully organized by birth, and myelination of the axons in these regions is quite complete. Shortly after birth, myelination of axons in the cerebellum (concerned with motor coordination) and midbrain begins, and thereafter—by the end of the first year or early in the second year—it begins in various parts of the forebrain, including the cerebral cortex.

The last brain structure to mature is the cerebral cortex, the seat of higher mental functions, including perception, memory, judgment, and reasoning, but here also maturation of all areas of the cortex does not occur simultaneously. Those cortical areas concerned with sensory processing mature earliest, followed by motor areas. But those areas concerned with the more sophisticated aspects of brain function—the so-called higher-order association areas of the brain, concerned with planning, intentionality, and other aspects of one's personality—are still myelinating axons and rearranging synapses up until the age of 18 or so! This includes much of the frontal lobes as well as parts of the temporal and parietal lobes of the cortex.

Recent imaging studies have extended our understanding of brain and cortical maturation. Studies have been carried out on children between the ages of five days and 15 years, using positron emission tomography (PET) scanning to determine glucose utilization by various parts of the brain. Glucose is the primary energy source for neurons (and other cells); the more active a neuron is, the more glucose it uses. It was observed that glucose utilization in newborns is largely limited to the brain stem, parts of the cerebellum, and certain subcortical structures. Very little glucose utilization was observed in the cortex itself, indicating relatively low neuronal activity there. By two or three months of age, glucose utilization increases significantly in some cortical areas, especially in the occipital cortex, which is involved in visual processing and perception. Not until six to eight months is significant activity observed in the frontal lobes, and again some parts of the frontal lobes show more activity than others.

Glucose utilization by the brain increases through early childhood and, interestingly, it peaks between four and seven years of age (depending on brain region), at which point glucose utilization is about twice the level of that in the adult brain. Glucose utilization by the brain then slowly subsides to adult levels through childhood and adolescence. The peaking of glucose utilization by the brain at four to seven years of age perhaps relates to the enormous synaptic plasticity of the brain at these early ages. Many new synapses are being formed, others eliminated, and synaptic circuits refined; but now I am getting ahead of the story. We shall come back to these issues later.

Mechanisms Underlying Brain Development

Let us return to the earliest stages of nervous system development and consider what is known about the underlying biological mechanisms. All neural tissue derives from neural plate cells, as shown in Figure 1-1, but what causes these cells on the dorsal side of the very young embryo to become neural plate cells? Two layers of cells initially make up the very young embryo: ectodermal cells, which cover the surface of the embryo and will eventually form mainly skin, and endodermal cells, which line the embryo and will form the digestive system and internal organs. A third layer of cells—the mesoderm, which will become muscle, bone, and connective tissue—develops next, and during its formation it migrates between the ectodermal and endodermal layers, initially on the dorsal side of the embryo. It turns out that the migrating mesoderm provides the signal for ectodermal cells along the dorsal surface of the embryo to become neural plate cells.

This aspect of brain development was first shown by two German biologists, Hans Spemann and his student, Hilde Mangold, who, in the 1920s, took mesodermal cells from the dorsal part of the salamander embryo and transplanted them to other parts of the embryo. They found that the transplanted mesodermal cells were capable of inducing any ectodermal cells to become neural plate cells, not just those on the dorsal part of the embryo. Thus, if they transplanted mesodermal cells from one embryo into another, they could induce the formation of two neural plates and, in some cases, the development of two nervous systems in the animal. Conversely, if they prevented mesodermal cells from migrating underneath the ectoderm in the early stages of embryonic development, no neural plate formed and the embryo lacked a nervous system.

How does the mesoderm cause ectodermal cells to become neural plate cells? It has long been suspected that a chemical signal from the mesoderm mediates this induction. For example, if pieces of embryonic ectoderm are cultured in the presence of mesoderm, they will become neural plate cells, but in the absence of mesodermal cells they will not. By placing porous filters between ectodermal and mesodermal layers, it is possible to define the size of the signal molecules, and these experiments suggest that the inducers are small proteins. If the porous filters are too small to allow small proteins to pass through, the ectodermal cells fail to become neural plate cells.

Three small proteins have now been shown to be neural plate inducers in amphibians: noggin, chordin, and follistatin. All of these proteins are thought to be secreted by mesodermal cells. Curiously, they act by binding to another secreted protein, called

BMP, and preventing it from interacting with ectodermal cells which would cause the cells to develop into skin. Thus, it is not the inducers acting directly on the ectodermal cells that causes them to become neural plate cells, but the lack of stimulation by BMP that results in their becoming neural plate cells. Why it works this way is not clear, but this first stage of brain development illustrates a principle seen again and again in brain growth, maturation, and function; namely, the key role of chemical signaling between cells.

Proliferation of Neural Cells

I noted earlier that as many as half a million cells might be generated per minute, on average, for the first four months of gestation in humans. How and when does this happen? Proliferation begins upon closure of the neural tube and initially takes place almost exclusively around its inner surface—an area called the germinal zone. Initially, the neural tube is just one or a few cell layers thick, but it rapidly thickens, enlarging from the inside out. Dividing cells undergo characteristic movements, shown schematically in Figure 1-5.

A cell getting ready to divide is bipolar in shape with a branch extending to the inner surface of the neural tube and another to

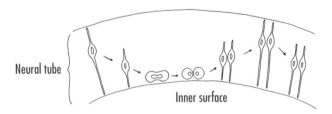

FIGURE 1-5 The generation of new cells along the inner surface (germinal zone) of the neural tube. The nucleus of a dividing cell moves up and down during the proliferation process. DNA synthesis occurs when the nucleus is away from the inner surface; cell division occurs along the inner surface.

or toward the outer surface of the tube. The nucleus of a dividing cell is situated away from the inner surface of the tube while it synthesizes DNA, but then it migrates to the inner surface of the neural tube. The cell withdraws its branches, rounds up, and divides. After cell division, the two daughter cells extend new branches, the nuclei migrate deeper into the tube, and the process repeats.

The genesis of cells destined to become neurons begins as soon as the neural tube forms (at three weeks of development) and reaches a peak in the seventh week, but then is largely completed by 18 weeks. Some neurons are generated later in the fetal stage of life, and some even in the first few postnatal months, but most neurons are generated by just four months of gestation. Glial cells, on the other hand, are generated continuously throughout gestation and even throughout life, though at a low rate.

The proliferation of the neural progenitor cells is under the control both of extrinsic growth factors—specialized proteins that act directly on cells to promote their division—and intrinsic factors—intracellular mechanisms that limit cell division. Cells exit the cell cycle—stop dividing—when the negative signals exceed the positive ones, but what the various signals are and how they are controlled are still poorly understood.

One idea that has been proposed to explain how cells leave the cell cycle is that there is a mechanism leading to asymmetric cell division at some point during proliferation. As long as cells divide symmetrically, they continue to generate more precursor cells. On the other hand, in asymmetrical cell division, one of the cells has less of a particular molecule than the other, causing the cell to leave the cell cycle. Is there any evidence for such a molecule? Recent experiments in mice, in which a protein called β-catenin is altered genetically so that it is more resistant to degradation, resulted in precursor cells excessively reentering the cell cycle rather than leaving it during early brain development. These mice grew grossly enlarged brains. Particularly striking in these animals is the cortex, which developed deep folds in the modified animals. In normal mice, on the other hand, the cortex is quite smooth, reflecting a much smaller structure. Unchecked neural cell proliferation can also lead to cancers—neuroblastomas—and this happens in the young developing brain. (See Chapter 5 for more discussion of cell proliferation.)

When cells exit the cell cycle, they typically move away from the germinal zone and form a distinct layer distally—called the intermediate zone. Cells in the intermediate zone are mainly young neurons that will never divide again. Where they will reside in the brain and even what kind of neuron they are likely to become are now essentially established. Some cells that migrate from the germinal zone retain the ability to divide, and a number of these cells form important brain structures, including the basal ganglia—subcortical areas that are involved in the initiation of movement. Certain cerebellar cells also proliferate after migration away from the germinal zone, and neural crest cells often divide after they have reached their final destination.

In cold-blooded vertebrates, such as frogs or fish, proliferative cells remain in the adult brain and continue to divide and generate new neurons. A particularly clear example is the retina of fish, which continuously adds neurons during the animal's life. In other words, the retina continues to grow as the animal grows over its life span. But most neuroscientists believe this is the exception; in most species, especially mammals, new neurons are not often generated in the adult brain.

Recent research has identified germinal cells in at least two regions of the mammalian brain, one is the hippocampus, a region of the brain concerned with the long-term storage of memories. There is some evidence that these germinal cells in the hippocampus can generate new neurons, but whether these new neurons become incorporated in the neural circuitry of the hippocampus is as yet uncertain. The nature and significance of other proliferating cells in the adult mammalian brain are also unclear and the subject of much controversy at present. Some investigators believe that many of these proliferating cells are glial cells. We shall return to this important issue in Chapter 5.

Migration of Young Neurons

From the intermediate zone, the young neurons must migrate, often considerable distances, to take up their final position. How this happens varies from region to region. In some parts of the brain, such as the retina and spinal cord, cells migrate in response to chemical clues, both positive and negative, present in the area. In other parts of the brain, such as the cortex and cerebellum, specialized glial cells, called radial glial cells, provide a scaffolding along which the neurons migrate. The cell bodies of these glial cells reside in the germinal zone, but they extend a branch to the surface of the brain as shown in Figure 1-6.

Electron microscopy has shown in the intact brain that migrating neurons are entwined around radial glial cell branches, and in tissue culture, neurons have been observed migrating along radial glial cell branches. In a mouse mutant that has a cerebellar defect in which the radial glial cells degenerate early, many of the cerebellar neurons do not end up in their proper positions and the animals show severe movement deficits. In normal mice (and other animals) the radial glial cells remain until neuronal migration is complete and then they disappear.

The sequence in which the cells migrate in the developing brain varies among different brain regions. In the cerebral cortex, for example, the first neurons to complete cell division and to migrate form the deepest layer of the cortex (so-called layer 6). Cells that proliferate later form the more superficial layers. In other words, the cortex grows from inside out. In the retina the opposite happens. The first cells generated (the ganglion cells) migrate across the retina where they take up residence, and cells generated later form layers of the retina closer to the germinal zone.

Differentiation of Neurons

Once the young neurons arrive at their final destination, they are first specified. That is, the kind of neuron they will become is

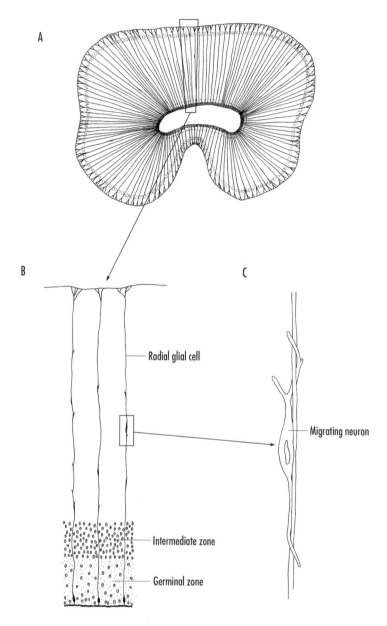

FIGURE 1-6 A: Radial glial cells in the developing cortex of the brain.
B: The radial glial cells extend from the germinal zone on the inner
surface of the neural tube to its outer surface.
C: Neurons from the intermediate zone migrate along the processes of
the radial glial cells to find their proper position in the brain.

determined. They next undergo differentiation: they extend branches characteristic of the type of neuron they are and begin to make synaptic contacts.

Almost needless to say, neurons are very complex cells of various sizes and extending numerous branches that often extend considerable distances as seen in Figure 1-4. Furthermore, each neuronal cell type usually has a unique branching pattern. What mechanisms underlie the specification and differentiation of neurons? Again, both intrinsic and extrinsic factors are at play. Specific types of neurons tend to be generated at specific times during development, and often all neurons of that type differentiate more or less simultaneously. In the retina, for example, the ganglion cells are specified and differentiate first, followed by cone photoreceptors, amacrine, and horizontal cells. The rod photoreceptors and bipolar cells differentiate last.

What triggers the specification and differentiation of a precursor neuron into a particular cell type? The local environment—the chemical signals the cells encounter—is clearly critical and this depends on the cells' position in the tissue. In other words, signals from nearby cells determine a cell's fate. Thus, extrinsic signals are key in the process. However, over time, the options for a cell to become a particular type of neuron are limited. That is, a precursor neuron is receptive to a specific inducing signal for only a particular window of time. Thus, intrinsic mechanisms are also at play in neuronal specification and differentiation and are also clearly important. To summarize, to become a particular type of neuron, a precursor cell must be in the right place at the right time during development.

In some species, especially invertebrates, in which there is little cell migration during development, intrinsic mechanisms play the major role in neuronal specification and differentiation. That is, the type of neuron a precursor cell becomes is determined by inherited developmental directives—the cell's lineage determines its fate. Thus, if a precursor cell is destroyed during development, the organism develops without the cells that it was to become. Other precursor cells cannot substitute for the deleted cell.

However, induction by position-dependent signals is clearly the major mechanism by which precursor neurons become specified and differentiate, especially in the vertebrate brain. Studies on the developing eye of the fruit fly, *Drosophila melanogaster*, have provided important insights into how this happens and the molecular mechanisms involved. The fruit fly's eye is a mosaic eye, consisting of individual photoreceptor units, called ommatidia. Each ommatidium has eight photoreceptor cells that can be individually identified and that are precisely arranged in the structure. During development, R8 (R stands for "retinular cells," the technical name for these photoreceptor cells) differentiates first. This is followed by R2 and R6, which differentiate together, then R3 and R4, followed by R1 and R5. The last cell to differentiate is R7, which contains a molecule that senses wavelengths of light in the ultraviolet region of the spectrum. The other photoreceptor cells respond to light in the visible spectral region (visible to us, that is).

The strict sequence of development of an ommatidium in the fruit fly eye suggests that earlier cells are responsible for the differentiation of the later cells with R8 taking the lead role. This was learned from experiments in which the early developmental sequence was disturbed. In such cases, the entire ommatidium did not form properly. Even more revealing were experiments on a mutated fly in which the R7 photoreceptor did not form at all—discovered because these flies do not respond to ultraviolet light. This mutant, called "sevenless," has told us much about the nature of the signaling mechanism and how the cell responds to such extrinsic signals.

The mutated gene in the sevenless fly codes for a protein that extends across the membrane of the cell as shown in Figure 1-7. On the outside of the cell, the protein serves as the receptor for the signal that tells the cell to become an R7 cell. On the inside of the cell, it acts as an enzyme when the receptor part of the molecule is activated. The enzyme part of the protein is a kinase, an enzyme that adds phosphate groups to proteins. Adding phosphate groups typically changes proteins' properties; activating or

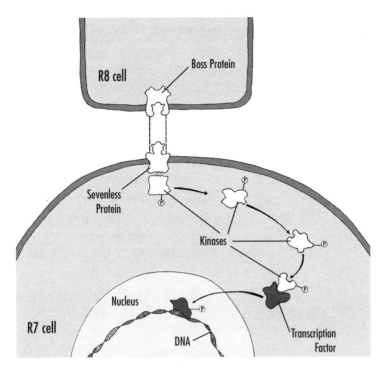

FIGURE 1-7 The interaction of the R8 cell with a precursor cell, and the sequence of events that leads to the differentiation of the precursor cell into an R7 receptor cell. When the boss protein on the R8 cell binds to the sevenless protein on the precursor cell, the sevenless protein acts as a kinase, adding phosphate groups to intracellular proteins. The intracellular proteins might themselves act as kinases, adding phosphate groups to other proteins and activating them. Eventually, proteins, called transcription factors, are phosphorylated. These proteins migrate into the nucleus, bind directly to DNA, and regulate gene expression, thus leading to the differentiation of the cell.

inactivating them if, for example, they are enzymes. Once the receptor part of the sevenless protein activates the intracellular kinase activity of the molecule, a series of biochemical events is initiated that ultimately leads to differentiation of the cell. If the cascade is not initiated—that is, the receptor protein is defective as in the sevenless mutant—the cell follows a default pathway and becomes a nonneural cell. In other words, it differentiates by way of intrinsic directives, but not into an R7 cell or even into a photoreceptor.

What is known about the signal that activates the sevenless protein? A second mutation in fruit flies in which the developing ommatidium also fails to form an R7 cell has provided an answer. This mutant is called "bride of sevenless" or "boss" and it affects the R8 cell. The defective gene codes for a membrane protein that is found in R8 cells in normal flies. We think that part of this protein extends from the surface of R8 and activates the sevenless receptor on the R7 precursor cells; this membrane protein is the signal (see Figure 1-7). Thus, direct contact between the cells is required for activation of the sevenless protein.

Scientists have also made progress in understanding what happens in the R7 precursor cell after the sevenless receptor protein has been activated. A number of downstream proteins have been identified and many are kinases themselves. Thus, activation of the sevenless kinase leads to activation of a number of other kinase enzymes. Targets of at least some of these kinases turn out to be transcription factors—proteins that migrate into the nucleus of the cell and turn on or off the expression of various genes by interacting directly with the regulatory regions of genes, the so-called promoter regions. The idea is this: when the sevenless receptor-kinase protein is activated by the R8 cell, a number of other kinases and transcription factors are eventually activated that lead to the expression of the appropriate genes needed to turn a precursor cell into a R7 photoreceptor cell.

Scientists believe that in other situations diffusible substances released by nearby cells control the differentiation of precursor cells, but that the same principles as described above apply. Some of these signaling molecules have been identified as small proteins. Some of the so-called growth factors described earlier play this role. These proteins all activate membrane receptors linked to a cascade of intracellular kinases that ultimately turn on or off specific genes. Thus, the general scheme shown in Figure 1-7 probably holds for the differentiation of neurons and glial cells throughout the brain.

How Do Axons Find Their Way?

Once neurons are specified and begin to differentiate, they extend both dendritic and axonal branches. This leads to the formation of synapses between neurons and, ultimately, to the wiring of the brain. A critical question is how axons find their way to their targets and how they know which cells to synapse upon. Often axons must travel considerable distances to reach their target neurons.

Again, chemical signaling is implicated as playing a critical role in cell-cell recognition. The notion is that as neurons differentiate, they become chemically specified; they make specific membrane proteins that extend from their surface and enable innervating axons to recognize them. Experiments that support this chemoaffinity hypothesis date back a century, but it was work carried out in the early 1940s by Roger Sperry at the University of Chicago that established the idea. Sperry studied the projection of retinal ganglion cells to a brain region called the tectum in animals such as fish or frogs. In these cold-blooded animals the tectum is the major target for the ganglion cell axons. The projections from retina to tectum are orderly; ganglion cell axons from one part of the retina innervate a specific region of the tectum. The projections are called topographic and are quite invariant from one animal to another. In other words, a retinal map is impressed on the tectum. The right retina projects to the left tectum and vice versa, and the tectal map is inverted relative to the retinal map, as shown in Figure 1-8.

In cold-blooded vertebrates such as fish and frogs, central nervous system (CNS) axons regenerate after they are severed and remake synaptic contacts (this, unfortunately, does not happen in mammals; see Chapter 5). Sperry took advantage of the regeneration of CNS axons in fish and frogs to show first that, if you sever the optic nerve in a newt (a frog-like animal), the axons regenerate and reform synaptic connections in the tectum. Furthermore, their vision is restored. How specific are these new connections? To test this idea, Sperry severed the optic nerves in frogs, but then

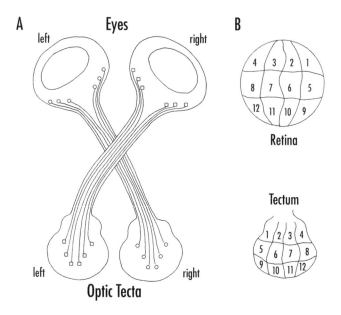

FIGURE 1-8 A: Ganglion cell axons from one eye of a fish or frog project in an orderly fashion to the tectum on the opposite side of the brain. B: Specific regions of the retina (indicated by numbers) project to specific regions of the tectum, forming what is called a topographic map. The tectal map is inverted relative to the retinal map.

rotated the eyes 180° before reattaching them in the socket. He found that after the optic nerves had regenerated, the animals could see once again, but their visual world was upside down and inverted from right to left! When the animals were feeding, they consistently misdirected their attempts to capture their prey by 180°. If prey they wished to capture was up and to the right, they moved down and to the left, and vice versa.

These experiments clearly showed that the severed ganglion cell axons had grown back to reinnervate the neurons with which they were originally connected. But because the animals' eyes were inverted, they saw an inverted world and responded in this way. Over time, there was no recovery; the animals were permanently altered (again, mammals are different in this respect; see Chapter 5). Sperry concluded that optic nerve axons can recognize the cells they are intended to synapse upon; that is, the cells have complementary markers that allow for recognition.

Direct evidence for recognition between retinal and tectal cells has come from the work of Friedrich Bonhoeffer and his colleagues in Germany. They took pieces of retina and tectum and dissociated the cells. By marking cells from the dorsal and ventral regions of the retina and tectum, they showed that cultured cells from the ventral part of the retina adhere preferentially to dorsal tectal cells and vice versa. In other words, the cells could recognize each other regardless of which part of the retina or tectum the cells came from.

How specific is this recognition? Present evidence indicates it is not cell specific, but region specific. That is, the retinal axons are not strictly wired to specific tectal cells. Rather, the ganglion cell axons have a strong affinity for tectal cells from a particular region and they will make connections with cells in that area. If, for example, ganglion cell axons from the chick retina are allowed to grow in a tissue culture dish that is coated with alternating stripes of cell membranes from either the anterior or posterior part of the tectum, axons from the anterior part of the retina (which normally innervate the posterior part of the tectum) in the chick will grow only on the stripes made up of posterior cell membranes, and vice versa. Furthermore, the axons actively avoid the inappropriate cell membrane stripes. From these experiments (and others) has come the realization that there are positive and negative recognition factors at work in this process. That is, some factors attract axons whereas others repel them.

A number of these factors have been identified as small proteins, but almost certainly all of the factors are not yet identified. The assumption is that these cell adhesion or repulsion proteins interact with receptors on the cells and or axons and activate intracellular enzyme cascades similar to those described earlier between the R8 and R7 cells in the developing fruit fly's eye. Activation of certain receptors and cascades leads to synapse formation between two cells; activation of other receptors and cascades tells the axons to go elsewhere.

Axons grow by way of specialized structures called growth cones that are flattened expansions at the ends of growing axons. A characteristic of growth cones is a prominent array of fine processes that extend from the growth cone as shown in Figure 1-9.

While axons are growing and seeking their way, the growth cone is in constant motion, extending and retracting its fine processes as it explores the area surrounding it. As the growth cone moves along, it adds new membrane to the axon, lengthening it.

The rate and direction of growth cone movement are affected by many factors. The presence of substances in the environment that attract or repel the axons is one key element, and long ago Sperry suggested that a gradient of a chemical substance could

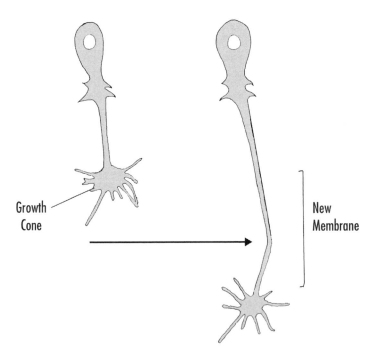

FIGURE 1-9 *How an axon grows. The growth cone at the end of a growing axon extends fine processes that explore the surrounding area. The axon adds new membrane near the growth cone, thus extending its length.*

guide axons to their target areas. Then cell-to-cell recognition mechanisms would promote synapse formation. He pointed out that two or more chemical gradients at different angles could provide growing axons with quite specific positional information as they grow. But other factors also contribute to growth cone movements, including the substrate on which the growth cone is moving, and even electrical fields in the growth cone's vicinity. Texture and adhesiveness of substrate, as well as the presence of recognition factors in the substrate, appear to be important.

When axons are required to grow long distances to reach their targets, two other mechanisms have been proposed to play a role. In some situations, guidepost neurons found at intermediate distances along the path the axons follow have been identified. Scientists think that the guidepost neurons provide a secreted attractant chemical that is sensed by the growth cone. Axons grow toward the guidepost neuron, but do not stop when they reach the cell. Rather, they move on to the next guidepost neurons, perhaps because there are repulsion factors on the surface of the guidepost neuron itself. As would be expected, damage to a guidepost neuron can disrupt axonal pathfinding.

A second proposed mechanism depends on the fact that initial axonal pathfinding in the brain occurs early in development, when distances between structures are much shorter than they are later in development or in the adult. Thus, early pioneer axons might serve to mark the path for axons coming along later, or axons migrating later might simply grow along the surfaces of earlier axons.

Synapse Formation

When axons reach their target, they make synaptic contacts with those neurons they recognize. Synapse formation requires reciprocal interaction between growth cone and cell to be innervated. Substances released from the growth cone initiate the formation of postsynaptic structures; conversely, the postsynaptic element provides signals to the growth cone to develop into a mature synapse. Much of the information we have concerning synapse formation has come from study of the innervation of muscle by the axons of motor neurons. This synapse, called the neuromuscular junction (NMJ), is large, employing acetylcholine (ACh) as its transmitter. When a motor neuron axon is active, it releases ACh from its terminals and the ACh diffuses to the muscle membrane, where it interacts with specific proteins that form channels through the membrane. When activated by ACh, these channels allow charged ions to cross the membrane. The movement of charged ions across the membrane results in voltage changes across the muscle membrane, initiating contraction of the muscle. Synapses in the brain work basically the same way, except that the voltage changes that occur across neuronal cell membranes result in the generation of electrical signals that move along axons (rather than contraction as in muscle cells).

The first clues concerning a trophic interaction between a motor axon synapsing on a muscle cell and the muscle itself came from studies on the distribution of the ACh channels along the muscle cell membrane. If ACh is squirted onto a normal adult muscle cell, the sensitivity of the cell to ACh is confined to the synaptic area; that is, the ACh channels are clustered at the synapse. However, if the axon is removed from the muscle—the muscle is deinnervated—the muscle becomes sensitive to ACh all over its surface. That is, in response to the deinnervation, new ACh receptors are synthesized by the muscle cell and they spread all over the cell's membrane.

Motor axons will reinnervate deinnervated muscle fibers in some animals and when this occurs, the sensitivity of the cell to ACh becomes confined once again to the synaptic region. Thus, reinnervation results in a decreased synthesis of ACh receptors and a clustering of the receptors to the synaptic site area. The same thing happens during development. Initially a muscle cell is sensitive to ACh all over its surface in the young embryo, but once it is innervated by a motor axon, the ACh receptors become clustered to the synaptic region.

These observations suggest that the motor axon terminal releases a substance that causes clustering of ACh receptors. Such a protein has been identified and is called agrin. This protein interacts with agrin receptors on the muscle membrane, which initiates a series of intracellular biochemical reactions resulting in phosphorylation of proteins. So once again, a chain of biochemical reactions similar to that shown in Figure 1-7 is implicated as playing a key role.

Agrin almost certainly contributes more to synapse formation than simply clustering the ACh receptors, but it also is not the whole story. Another protein called ARIA (acetylcholine receptor inducing activity) was isolated from chicken brain and causes an increase in ACh receptor synthesis. Again, it is thought that ARIA is released from the motor axon nerve terminals.

The story is by no means complete and, as noted above, it is believed that chemical signals also go from the postsynaptic cell to the innervating presynaptic axon, and these signals trigger changes in the presynaptic terminal that lead to the formation of a mature synapse. But what these signals are is not known.

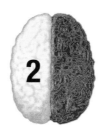

MATURING A BRAIN

In the first chapter I described how the brain is formed—how precursor cells become specified as neurons, how these nascent neurons migrate to appropriate positions in the embryonic brain and then differentiate into specific neuronal subtypes. I discussed how axons find their way to their targets and, finally, how synapses are formed. These events establish the framework of the brain, and clearly they depend on genetically specified molecular mechanisms. Thus, what I have described so far depends mainly on nature.

Next, I'll discuss the maturation of the brain, brain circuitry, and behavior, and here experience—nurture—plays a critical role. How much of a role remains contentious—the nature-nurture debate is certainly not settled, and perhaps no scientific debate of the last century has generated more controversy. Everyone seems to know the answer and reams have been written on the question, almost always with a definite point of view. But because we do not know the final answer, these views come under heavy attack from those on the other side of the argument.

Recent studies on the maturing brain provide us with new ways of thinking about the issue and I will emphasize them in this chapter. I begin by noting several important themes with regard to neuronal, synapse, and circuitry maturation. First, neurons are initially overproduced in many parts of the brain and a significant amount of cell death is a long-recognized feature of brain maturation. Much of this cell death appears to involve competition for synaptic sites. Neurons whose axons form synapses survive—they are the winners. Those that fail to find synaptic targets die—they are the losers. But neuronal death is reciprocal—postsynaptic cells depend on having synapses made upon them. For example, if input to a brain region is removed, excessive neuronal death is observed there. Conversely, if a structure receives excess synaptic input, more neurons might survive in it than ordinarily. The extent of neuronal cell death varies in different brain regions. The cerebral cortex appears to undergo only a modest amount of neuronal degeneration during development, whereas the spinal cord and some regions of the hindbrain might lose 30-75 percent of their neurons during maturation of the nervous system.

A second important theme of brain maturation involves a restriction of axonal terminal fields and a rearrangement and refinement of synapses. Newly formed neurons typically extend their axonal branches over a wider area than they do in the mature nervous system, and they make synapses upon more cells than they do in the adult brain. Thus, during brain maturation some axonal branches are lost whereas others are formed, and some synapses are lost while others are being made. In other words, neurons initially establish qualitatively appropriate connections during brain development, but then during maturation the connections are rearranged and refined to provide the more precise relationships found in the adult brain. Well-studied examples include the innervation of muscles and neurons of the autonomic system that regulates our internal organs. At birth, axons typically innervate several muscle fibers or autonomic ganglion cell neurons as shown in Figure 2-1.

Newborn

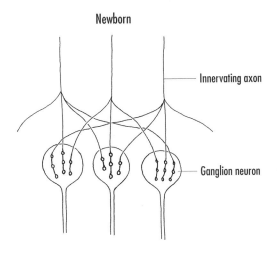

Innervating axon

Ganglion neuron

Adult

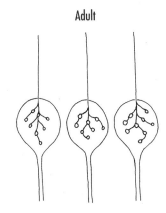

FIGURE 2-1 *A schematic view of the rearrangement of synapses that occurs during maturation of an autonomic ganglion. Initially, axons innervating the ganglion extend branches to several ganglion cells and make synapses on all the contacted cells. In the adult brain, an axon typically contacts only a single ganglion cell, but that axon now forms many synapses with the cell it innervates.*

Over the first few weeks of life, the innervation of the muscle fibers and ganglion cells changes so that eventually only one axon innervates a muscle fiber or autonomic ganglion cell. However, the axonal branch innervating the muscle fiber usually makes more synapses on its target. Thus, during this process, certain axonal branches and synapses are lost, but then new branches and synapses are made, so that there might be more innervation of a muscle fiber or ganglion cell. Further, the innervation is now more specific and stronger as shown in Figure 2-1, resulting, presumably, in better and more-refined neural control of the muscle or ganglion cell.

Mechanisms of Trophic Interactions

What mechanisms underlie the loss of cells, retraction of processes, and rearrangement of synapses during brain maturation? A clear implication of these phenomena is that an exchange of information takes place between input neurons and their target cells, and this exchange regulates neuronal shape, connectivity, and even cell survival. It is no great stretch to suggest that this exchange of information is chemically mediated.

A number of substances released at synapses have trophic effects on postsynaptic neurons, causing them to extend or retract their branches. The most important substances in this regard are proteins called growth factors, examples of which are the neurotrophins, which regulate cell death, dendritic and axonal branching, and the extent and pattern of synaptic innervation in a brain region. Both presynaptic and postsynaptic neurons, as well as some glial cells, are known to release neurotrophins. Neurotrophins interact with specific membrane proteins, called Trk (tyrosine kinase-containing) receptors. These receptor proteins extend across the membrane with the portion on the outside of the cell available for neurotrophin binding, whereas the part inside the cell acts as a kinase enzyme. The binding of a neurotrophin molecule to a Trk receptor protein activates the kinase and initiates a series of intracellular biochemical reactions. These biochemical events within the cell can alter enzyme activity, gene expression, or whatever, by way of mechanisms like that described in the last chapter for the differentiation of the R7 photoreceptor in the fruit fly's eye and illustrated in Figure 1-7.

The first neurotrophin discovered—and still the best characterized—is nerve growth factor discovered in the early 1950s by Rita Levi-Montalcini, then a young postdoctoral fellow from Italy working with Viktor Hamburger, a developmental biologist, at Washington University in St. Louis. They were studying a phenomenon originally observed by Hamburger—that an excessive number of neurons die in the spinal cord of chick embryo following the removal of a nearby developing limb known as a limb bud. Although it was well known that some cell death occurred in the spinal cord during normal development, a surprisingly large number of neurons died following excision of the growing limb. They surmised that the target cells in the limb bud send a chemical signal to the innervating spinal cord neurons, which permits the neurons to survive. They further reasoned that the amount of the substance is limited and that this is why some of the neurons routinely die. Following loss of the limb bud, much less of the substance is available and massive cell death in the spinal cord occurs.

What is this chemical signal? Levi-Montalcini, working with a biochemist colleague, Stanley Cohen, at Washington University, soon isolated a fairly large protein, which they named nerve growth factor (NGF). NGF stimulated the survival and growth of spinal cord neurons and turned out to be the chemical signal released from the limb. For this research, Levi-Montalcini and Cohen were awarded a Nobel Prize in 1986.

NGF does not work on all neurons, but since the discovery of NGF, a number of other related neurotrophin proteins, including a protein called brain-derived growth factor (BDNF) and two closely related proteins, neurotrophin 3 and neurotrophin 4/5, have been identified. These proteins differ in terms of the types of neurons they act upon and the effects they exert. However, they all act on Trk receptors that are present on the responsive cells and that are specific for a particular neurotrophin.

In the chick, NGF acts mainly on spinal cord sensory neurons and also on autonomic ganglion cell neurons (which lie alongside the spinal cord). During normal development, about one-third of these neurons die, but if excessive NGF is applied to the cord, many of the cells survive. Conversely, when NGF is inactivated by an antibody continuously administered to a chick, virtually all of the spinal cord sensory neurons and autonomic ganglion cell neurons die.

In addition to promoting the survival of neurons, NGF promotes the growth of dendrites and the formation of synapses by the spinal cord and autonomic ganglion cell neurons. Figure 2-2A illustrates this growth in newborn rats that were given NGF daily for two weeks. The neurons (ganglion cells) were injected with a dense staining marker, visualized under the microscope and drawn. The dendritic arbors of the cells from the treated animals were considerably larger and more complex than those of the control animals.

Figure 2-2B shows how NGF stimulates the extension and direction of axonal growth. Both in culture and in the intact animal, growing axons turn toward a source of NGF. Thus, NGF seems capable of guiding axons. In the experiment shown in Figure 2-2B, a micropipette containing NGF was slowly moved around the culture disk. The growing axon elongated and turned in response to the NGF that was slowly diffusing out of the pipette.

Visual System Development

So far I have suggested that initially the brain is wired up in a qualitatively appropriate fashion as a result of intrinsic mechanisms. No experience is needed for this to happen. How good is this initial wiring? Electrical recordings of the neural activity generated by neurons in the primary visual cortex of newborn cats and monkeys by David Hubel and Torsten Wiesel, first at Johns Hopkins University in the 1960s and later at Harvard Medical School, are revealing in this regard. It is in this region that visual information is first processed in the cortex, and Hubel and Wiesel

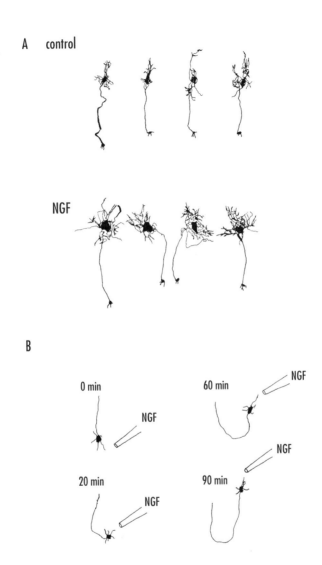

FIGURE 2-2 *The effect of NGF on the formation of dendrites (A) and axonal growth (B). NGF increases significantly the extent and number of dendritic branches on a ganglion cell neuron and it alters the direction of growth of an axon growing in a culture dish.*

showed earlier how visual input is analyzed by the cortical neurons. For example, neurons close to the input layers in the cortex
are highly orientation selective; that is, these neurons respond
best when a bar or edge of light with a specific orientation is
present in the appropriate part of the visual field. Visual neurons
all along the visual system typically respond to stimuli in a restricted region of the visual field—termed the receptive field.
What is striking in this experiment is that the neurons required
elongated visual stimuli that had a specific orientation. If the orientation of the bar or edge of light was skewed from optimal by
more than 10°, the cells responded less well. These first cortical
neurons are called simple cells as shown in Figure 2-3A.

Hubel and Wiesel found that farther away from the input layers of the cortex the cells had even stricter requirements if they
were to be activated maximally. Not only must the elongated
stimuli have a precise orientation, they must also be moving at
right angles to the direction of orientation. If the stimuli are not
moving, or are not properly oriented, the neurons respond less
well. These second cells are termed complex cells.

Even more specialized cells are also observed in this region of
the cortex, and these neurons appear to represent a third stage of
processing. Some of these specialized complex cells show direc-
tion-selective properties—that is, they respond well only to an
oriented bar of light moving in a specific direction as shown in
Figure 2-3B. Other cells (end-stopped cells) require a bar of specified length—they add yet another restriction to the stimulus
needed to best activate the cell.

The overall picture derived from these studies is that an enormous amount of neural processing occurs already in the primary
visual area of the cortex. Intricate synaptic connections between
neurons are obviously required for the establishment of cells with
such sophisticated responses. Therefore, this area is ideal to study
in newborn animals, which are visually inexperienced. Is the
neuronal machinery present at birth or does it develop in response
to the visual environment? It turns out that the answer is
interesting.

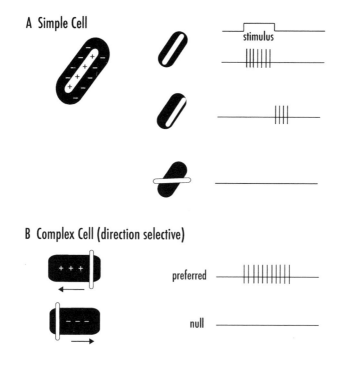

A Simple Cell

stimulus

B Complex Cell (direction selective)

preferred

null

FIGURE 2-3 *Receptive field maps (left) and responses (right) for simple (top) and complex (bottom) cortical neurons.*
A: The simple cell responds best to an oriented bar of light that fits its central excitatory zone (+ symbols). Moving the bar into the surrounding region (– symbols) elicits inhibition and an OFF response from the cell. Stimulating the field with an inappropriately oriented bar of light results in little or no response, the excitatory and inhibitory regions interact, canceling the cell's response.
B: The complex cell responds to an oriented bar of light moving at right angles to the bar's orientation. The cell illustrated is direction selective. Movement in the preferred direction elicits vigorous activity; in the null direction, no activity.

When the electrical activity of neurons is recorded from the primary visual cortex of both newborn cats and monkeys, the responses are remarkably adult-like. The neurons show good orientation sensitivity and, if they are complex cells, movement sensitivity. Some cells are directionally selective and others show the property of end-stopping. Overall, the cells are somewhat less active than adult cells and occasionally a cell is encountered that cannot be activated by visual stimuli or has poor orientation selectivity. But it is clear that this area of the cortex is essentially ready to go at birth—no experience is needed for the development of the sophisticated responses. Much of the requisite circuitry must be determined genetically—nature is all that is required.

Not everything is exactly adult-like in the cortex of newborn cats and monkeys. For example, input to the primary visual cortex is from a region of the brain called the thalamus, and specifically from neurons of a thalamic nucleus, the lateral geniculate nucleus (LGN). (A brain nucleus is a cluster of neurons involved in a specific neural function.) LGN neurons receive their input directly from the retinal ganglion cells; thus, most visual information reaches the cortex by way of these neurons. Visual information from the two eyes is obviously separate in the two optic nerves innervating the LGN and, interestingly, it is also kept separate in the LGN. Thus, the LGN neurons providing input to the primary visual cortex carry information from either the left or right eye, but not both; they are termed monocular.

The cortical neurons receiving direct input from the LGN cells are also monocular; they receive input from just one or the other eye. Further, they are clustered in columns or stripes that run laterally across the cortex. The columns are somewhat irregular as shown in Figure 2-4, but are about 0.5 mm wide. The stripes alternate so that one stripe has cells driven primarily by the right eye, the next stripe by the left eye, and so forth.

The ocular dominance columns can be visualized by injecting a radioactive amino acid into one eye and examining the pattern of radioactivity in the cortex (Figure 2-4). This works as follows: The radioactive amino acid is taken up by the retinal ganglion cells in the injected eye, made into protein, and then transported to the LGN by way of the ganglion cell axons. All axons have specialized transport mechanisms that move substances from the cell body where they are synthesized down the axons to the terminal synapses where they are needed. Typically, some of the ra-

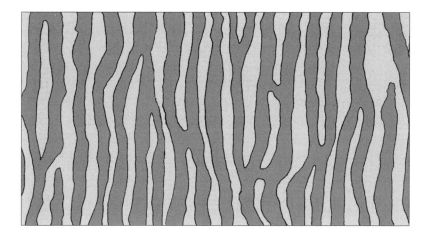

FIGURE 2-4 The anatomical demonstration of ocular dominance columns in the monkey cortex. Input from the two eyes alternates, forming regular bands—the ocular dominance columns—about 0.5 mm wide.

dioactive protein is released at the synapse, where it is taken up by the LGN neurons. The LGN neurons, in turn, transport some of it to the cortex by way of the axons of the LGN neurons. This takes about a week, at which point the LGN axon terminals that received input from the injected eye are radioactive. Sections cut along the cortex at the LGN input level demonstrate the ocular dominance columns, because radioactivity, like light, exposes silver grains in photographic film. Thus, a piece of film placed on the tissue section reveals the pattern of the radioactive axonal terminals.

Above and below the LGN input layer of the cortex, the neurons receive input from both eyes, the extent of which depends on how far away the neurons are from the input layer. Thus, most cortical neurons are binocular, although one eye usually tends to dominate the neuron. Near the input layer, as noted above, there are cells that receive all of their input from one eye or the other—they are monocular. In the newborn cat or monkey, on the other hand, it was discovered by recording from the cortical neurons

that all the cells in the cortex are binocular; there are no monocular cells. That is, the cortical cells near where the LGN axons
terminate, which normally receive input from just one eye, initially receive input from both eyes.

Anatomical studies in which a radioactive amino acid was
injected into one eye also showed no ocular dominance columns;
rather, radioactivity was spread evenly across the input layers of
the cortex. What is going on? Anatomical examination of single
innervating LGN axons provided the answer. The arborizations
of the axons innervating the cortex at birth are not segregated
into stripes but extend widely across the cortex. Only after a few
weeks do the axons retract and remodel their axonal terminal
fields to form the ocular dominance columns. This retraction process appears similar to that described earlier for the innervation
of muscle and autonomic ganglion cell neurons and as shown in
Figure 2-1.

Although the reshaping of the LGN axonal terminals in both
cat and monkey occurs postnatally, it does not appear to require
visual experience. Animals initially restricted in their visual experience develop perfectly normal ocular dominance columns.
Although visual experience is not required for postnatal development of the columns, neuronal activity is required. If, for example,
a drug is injected into an eye that prevents retinal ganglion cells
from generating the electrical signals that travel down axons, ocular dominance columns in the cortex do not form. Also, in the
LGN, segregation of right and left eye activity does not occur
when retinal ganglion cell activity is stilled. In the absence of
light stimulation of the eye, how is the electrical activity of the
ganglion cells generated? Interestingly, the retinal ganglion cells
are spontaneously active once they differentiate, and it is this
spontaneous activity that is required to refine connections in both
the LGN and primary visual cortex.

The spontaneous activity of the ganglion cells is not random,
but waves of activity that pass across the retina are generated.
These waves, lasting two to eight seconds occur at one- to two-
minute intervals and are limited in their domain. That is, one

wave of activity does not travel over the entire retina. The waves of spontaneous activity seem important not only for segregation of input into eye-specific layers, but also for the topographic projections in the visual system. For example, the formation of the precise topographic map of the retinal ganglion cells on the tectum in cold-blooded vertebrates (shown in Figure 1-8) requires neural activity, and the guess is that the correlated activity of adjacent ganglion cell axons generated during the waves is critical for the refinement of the map from an initially coarse projection. If fish or frogs are raised under strobe lights, which synchronize activity of the ganglion cells, refinement of the topographic map does not happen. Thus, the timing differences in the generation of electrical responses among nearby cells during a wave provide information as to the relative location of the cells, and this is critical for the generation of a precise map.

To summarize, the initial wiring of the visual system requires not only intrinsic mechanisms but also neural activity. However, none of this depends on visual experience. Development of the retina, LGN, and primary visual cortex clearly depends on nature. But what about higher visual centers that are concerned with more specific aspects of visual recognition and perception— does their development require visual experience?

Although less is known about the development of higher visual centers, some evidence on this issue has been obtained by study of a visual area found in the inferior temporal part of the cortex that in humans appears to be involved with face recognition. Patients with lesions in this area, caused by a stroke, for example, fail to recognize familiar faces, including those of spouses. When the spouse speaks, an affected patient instantly recognizes who it is; hence, this is clearly a visual perceptual deficit.

When electrical recordings are made from neurons in the same area in visually inexperienced monkeys as young as six weeks of age, the neurons respond to complex images, including faces, much as they do in adult animals. Thus, even quite specialized visual areas do not appear to require visual experience to be wired up to perform their assigned function. However, behavioral studies of monkeys employing complex visual recognition tasks indicate that this ability develops to adult levels relatively slowly, over the first year of life. Therefore, other brain areas involved in visual recognition tasks must be developing over this time and their development might be influenced by experience.

So when does visual experience play a role in the visual areas that have been studied? The answer is after the initial wiring takes place, and this has been elegantly shown by depriving young animals of various kinds of visual stimuli.

Visual Deprivation

In cats, visual experience appears to play no role in the maturation of the visual cortex up to approximately three weeks of age. Thereafter, profound changes in cortical physiology and anatomy, as well as visual performance, occur following visual deprivation. Deprivation can be of light—the animals are raised in the dark— or of form—one or both eyes are subject to lid closure, or a light-diffuser is applied to one or both eyes, so that light can reach the retina but no crisp images are formed. Somewhat different effects result from these and other types of visual perturbations, but all result in clear and persistent visual changes. In essentially all cases, visual acuity is severely reduced and if visual deprivation is restricted to one eye, there are striking changes in the binocularity of the visual system.

The bottom line is that the cortical visual circuitry is initially quite labile: It can be easily and substantially modified in the young animal by altered visual experience for a period of time. This occurs not only in cats, but also in monkeys and other animals. It holds also for humans who at birth have, for example, a cloudy lens in one eye or both eyes, or have a misaligned eye. Effective vision is lost from the affected eye or eyes—a condition known as amblyopia. The evidence is strong that the changes underlying this loss of effective vision occur mainly in the cortex; the retinas remain quite functional and normal, as do the LGN neurons, during various types of visual deprivation.

Perhaps most striking are the changes induced in the cortex when just one eye is deprived of form vision—monocular deprivation. In a typical experiment, the eyelid of one eye of a cat is shut by suturing in the first postnatal week and the eyelid is kept closed for three months or so and then opened. Recordings made from neurons in the retina, LGN, and primary visual cortex indicate that no major changes occur in the properties of the retinal and geniculate cells. However, profound changes are seen in the cortical cells, particularly with regard to the binocularity of the cells.

What is instantly clear in the deprived animal is that few cells in the cortex are binocular. The overwhelming majority of the cells recorded receive their input from the open eye and are monocular; they can be driven only by the open eye. The few cells that are driven by the closed eye or have preference for the closed eye are highly abnormal. They give weak responses, and often nonspecific responses—they typically have poor orientation selectivity and they respond sluggishly. Also, a number of the cells recorded are unresponsive to light stimuli.

That the great majority of recorded cells have input from the open eye suggests that the open eye has taken over cells that normally would have received most if not all of their input from the closed eye. In other words, the open eye's input now occupies more territory than the input from an eye in a normal animal. What do the ocular dominance columns look like in a monocularly deprived animal?

Physiological experiments by Torsten Wiesel and David Hubel showed that the ocular dominance columns change quite dramatically in size in a monocularly deprived animal. The amount of cortex receiving input from the open eye is greatly expanded, whereas the amount of cortex receiving input from the deprived eye is severely restricted. Anatomical studies confirm these observations as shown in Figure 2-5B for a monocularly deprived monkey.

In the experiment shown here, radioactive material was injected into the open eye and a section cut along the input layers of the cortex. Not only is the amount of cortex devoted to the open eye considerably enlarged (lighter areas) as compared to the

A   Normal   B   Deprived

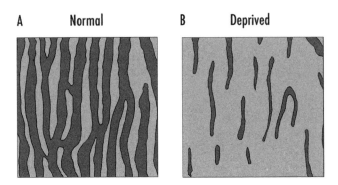

FIGURE 2-5 *The alteration in ocular dominance columns induced by monocular deprivation in a young monkey.*
A: The columns in a normal monkey.
B: The columns from a deprived animal. In the visually deprived monkey, the open eye input (light areas) occupies much more territory than the closed eye (dark bands).

amount devoted to the closed eye, but the columnar stripes reflecting input from the deprived eye are discontinuous. Compare this image with Figure 2-5A, which shows the ocular dominance columns in a normal monkey.

How might one eye take over cortical territory from the other? One plausible suggestion is that the axon terminals of the geniculate axons coming from the open eye do not retract as much as they do ordinarily, whereas the axon fibers coming from the closed eye retract much more. The implication here is that lateral geniculate axons compete for cortical space and synaptic connections in the young animal. As long as each eye provides the same input to the cortex, the competition is even, and both eyes end up having equal cortical representation. If the ocular input is not equivalent, the dominant eye ends up with more cortical space and presumably more cortical synapses. The notion of a competition for synaptic sites and territory was noted earlier in the discussion of cell death in the developing nervous system and appears to apply in many situations. We shall come back to this notion later.

Another possibility that might contribute to the open eye having more cortical representation than the closed eye is that its axons have more terminals. That is, in addition to an excessive retraction of axonal processes of geniculate neurons receiving input from the closed eye, the axon terminals receiving input from the open eye sprout new processes.

What happens if you visually deprive both eyes by suturing both eyelids shut or raising animals in the dark? In dark-raised cats, cells in the cortex are mainly binocular as in a nonvisually deprived animal, but the recorded cells are typically nonselective to orientation. In lid-sutured cats, binocular deprivation leads to many unresponsive cells or cells that respond erratically. Furthermore, cells in which it is possible to map receptive fields are mainly monocular—they seem to have lost their binocular connections. Loss of form vision thus causes somewhat different changes than dark-raising, and this is seen also in the recovery from the two forms of deprivation, as we shall see.

Pattern Deprivation

In the experiments described so far, all form vision was withheld from one or both eyes for a period, and severe defects were noted in the responses of the cortical neurons. It is possible to induce more subtle deficits in the responses of cortical neurons by restricting just one or another aspect of the visual world. One obvious experiment is to raise animals under conditions in which they are exposed to bars or lines of only a single orientation. When this is done, the neurons recorded from the cortex are biased with regard to the orientations to which they respond. A normal cortex has neurons that respond to all possible orientations; in animals raised in environments where they saw only horizontal or vertical stripes, the cells respond selectively to the orientations to which they were exposed.

Other experiments have extended this idea. If animals are restricted at an early age to environments in which they see little movement, or movement in only one direction, their cortical neurons seem to be either less movement sensitive or biased to movements to which they have been exposed.

Critical Periods

An important question is whether the human cortical circuitry can be modified throughout life. This is a somewhat difficult and also contentious question because some modifications to brain circuitry do occur throughout life, as will be described in Chapter 4. For example, we continue to learn regardless of our age and this learning causes molecular and probably structural changes in our brains. However, there is no question that the kinds of drastic changes to both cortical physiology and anatomy that occur as a result of visual deprivation in the young cat, monkey, or human do not generally occur in adults. In adult cats, monkeys, and humans, various kinds of visual deprivation even of long duration—months to years—do not have dramatic effects on visual performance, on the responses of cortical neurons, or on cortical anatomy.

To induce changes like those described above, the deprivation must occur when the animal is very young. The period of great susceptibility is called the critical period or sensitive period. In cats the critical period for the primary visual cortex begins at about three weeks and extends to four months; the period of greatest susceptibility for changes in ocular dominance columns peaks at about six weeks and then subsides. In monkeys, deprivation between birth and six weeks induces the most severe effects with a peak at about one month, but deprivation between six weeks and one year also causes deficits. In humans, deprivation between six months and six years of age causes amblyopia—severe loss of visual acuity—in the deprived eye.

During the periods of high susceptibility, short periods of visual deprivation can cause very severe changes. Indeed, just a few days' deprivation in the first two to four weeks of a monkey's life can result in changes about as severe as those seen in animals whose eyes are kept shut for several weeks later in the critical period.

The general notion of critical periods in cortical development has been questioned, because often there is neither a sharp start nor a sharp end to such sensitive periods. Some investigators believe, rather, that cortical modifiability is a continuum, with, at most, periods of more susceptibility. However, in the visual system, the notion of critical periods seems quite clear. On the other hand, we have also come to realize that there are different critical periods for different aspects of cortical and neuronal function. For example, in cats the susceptibility of directionally selective cells to alterations in their directionality is virtually over by six weeks of age, at the time when the susceptibility for changes in ocular dominance columns is at its peak.

In addition, critical periods can be modified by environment. Dark-raising alters the critical period for ocular dominance changes in two ways. The peak time of sensitivity is delayed from the norm of six weeks to about twelve weeks, and, second, the duration of the critical period is significantly lengthened. Dark-raising appears to slow down and even reverse maturation of the cortex. Up to three weeks of age, no differences in light- and dark-raised kittens are noted; however, after that the responsiveness of the cortical cells in dark-raised animals decreases. In addition, more non-orientation-sensitive cells are encountered in the dark-raised cats.

Interestingly, brief periods of light exposure in cats reverse the effects of dark-raising. As little as six hours of light exposure appears to shorten the critical period and to stimulate the maturation of the visual cortex. How might this come about? We don't know in detail, but if dark-raised cats are exposed to light for just a few minutes, changes in gene expression in the cortex can be measured within an hour or so. This indicates that even brief exposures to light in a dark-raised animal can induce biochemical changes in neurons of the cortex that affect their maturation.

Recovery

Changes induced in the cortex as a result of visual deprivation are difficult to reverse. For example, after monocular visual deprivation, little recovery is noted if nothing is done other than to open the closed eye, even if the eye is reopened during the critical period. In one experiment, the eyelid of a monkey was closed for nine days during the first two weeks of life. The eyelid was then opened and nothing further was done to the animal. At four years of age, recordings from the animal's cortex showed changes similar to those of a monkey who had one eye closed from birth to four years of age.

Substantial recovery can be induced, however, if the animal is forced to use the deprived eye, especially if this forced usage occurs during the critical period. If the deprived eye is opened and the formerly open eye closed, good recovery is observed. Ophthalmologists learned this trick long ago to treat children with amblyopia, having them wear a patch over the normal eye for periods during the day to improve the visual acuity mediated by the amblyopic eye.

On the other hand, although visual acuity in the deprived eye improves dramatically, binocular responses do not. When the responses of neurons in the cortex of cats treated like that are recorded, virtually all of the neurons turn out to be monocular, receiving input from one eye or the other, but not both. Thus, loss of binocularity in cortical neurons is not related to a loss of acuity. These two visual attributes can be quite independent of one another.

An independence of visual acuity and binocularity also results when the eyes are not aligned correctly, a condition known as strabismus. Thus eyes can turn out (wall-eyes) or turn in (cross-eyes). In wall-eyed people, vision usually alternates between the two eyes. When looking at objects to the right, they use the right eye and ignore the visual information coming from the left eye, and when looking left, they use the left eye. The visual acuity in both eyes is normal, but binocular interactions between the two eyes are lacking. In young animals who are made wall-eyed surgically, the cortical neurons are almost exclusively monocular; half respond to the left eye, the other half to the right eye, and virtually none to both eyes. In cross-eyed cats (and people) a monocular-deprivation-like deficit occurs. One eye—usually the straighter, becomes dominant and high-acuity vision is lost in the other eye. Most of the cells are driven monocularly by the dominant eye, and very few cells are binocular or driven exclusively by the crossed eye.

Enriched Visual Environments

The visual system is a most convenient part of the brain for studying the development and maturation of brain structures. Neurons along the visual system can be readily activated by presenting visual stimuli to the eyes, and activity of the neurons is easily recorded. Visual stimuli to an animal can be altered in various ways to explore the effects of environment on visual development. Several important conclusions from these studies have already been noted; others have not been emphasized so far. For example, although the primary visual cortex is very susceptible to significant alterations in its structure and function as a result of altered visual experience, neither the retina nor the LGN shows such drastic changes. Thus, not all brain regions are equally plastic—some are more hard-wired than others, and it appears that higher brain centers are more modifiable than lower ones. This conclusion extends to different kinds of animals as well. Cold-blooded vertebrate brains appear much more hard-wired than our brains, probably reflecting the fact that cold-blooded vertebrates have much less cortex relative to other brain structures as compared to mammals, and I shall return to this notion later.

A second point is that in all the experiments described so far, the animals were visually deprived in one way or another. This resulted in loss of visual acuity, binocularity of cortical neurons, orientation selectivity, movement sensitivity, and so forth. Furthermore, I earlier emphasized the notion of overproduction and pruning of neurons that occurs during maturation of the brain—restriction of dendritic fields, rearrangement of synapses, and even cell death. All of this might be summarized by the adage "Use it or lose it" as a key feature of brain maturation. What about going the other way? Can one enhance brain circuitry by, for example, providing an animal with an enriched environment?

Again, studies on the visual cortex have led the way in this regard, although it is also fair to say that the results are not nearly as clear-cut nor as convincing as those that have come from the deprivation studies. Nevertheless, we can reach some conclusions. The pioneering studies in this regard were carried out by Mark Rosensweig and his colleagues at the University of California, Berkeley, in the 1970s, who found that various regions of the cortex were heavier and thicker in rats reared from weaning in an enriched environment as compared to animals raised in a more impoverished environment. The enriched environment consisted of housing rats in groups of 10-14 in large cages with various play objects that were changed daily. In addition, these animals were put into a 4-foot square box containing other play objects for up to an hour a day. Impoverished environments consisted of animals being housed singly or in pairs in standard laboratory rat cages with no toys.

The biggest differences Rosenzweig and his fellow scientists found were those between rats housed in the enriched environments and rats housed singly, and changes in the occipital (visual) cortex were most prominent. They noted upon examining the tissue of the occipital cortex that the neurons were larger and the glial cells more plentiful in the enriched-environment rats.

Subsequently, William Greenough and his colleagues at the University of Illinois examined the structure and density of synapses in rats raised in an enriched environment. They found a small increase in the size of synapses—~8 percent—in animals raised from 25 to 55 days old in an enriched environment. Several other studies on rats confirmed this result, but a similar study in cats failed to find a significant difference. More interesting, and certainly more dramatic, was the finding that the number of synapses per neuron in the occipital cortex increased by 20-25 percent in rats raised under enriched conditions compared to those raised in an impoverished environment. The number of synapses per cubic millimeter didn't change much when the rats were raised in an enriched environment, but because the neurons are larger in these animals, and hence there are fewer of them per square millimeter, the number of synapses per neuron is greater. These results were confirmed in at least three studies, two in rats and one in cats.

When the individual neurons from the occipital cortex of enriched-environment rats were examined by methods that allow an examination of the cell's dendritic tree, the extent of the dendritic field was found to be increased by about 20 percent, a result congruent with the increase of synaptic number per neuron. The extent of the dendritic field was determined by measuring the total length of dendrite measurable in a tissue section. That is, this measure included both elongation of dendritic branches and the formation of new branches.

As noted, the studies described above were carried out mainly on rats between the ages of 25 and 55 days. When similar studies were carried out on older adult rats, most of the same results were found. The weight of the occipital cortex in the enriched-environment rats had increased, as had the number of synapses per neuron. Some question was raised as to whether the cortical dendritic fields of the older animals were as expanded as in the enriched-environment younger animals, but overall the results were surprisingly similar between young, adult, and middle-aged rats. The one effect not seen in the older rats was the small increase in synaptic size in the cortex of enriched-environment rats; that effect (if indeed real) seems limited to young animals.

The conclusions drawn from these studies is that enriched environments increase neuronal size, glial cell growth, synaptic density per neuron, and even synaptic size, but these effects occur in adult animals perhaps as readily as in young ones. Another caveat is that when animals (rats) are returned to the impoverished environment of standard laboratory cages, the enriched-environment synaptic growth regresses. Thus, the brain growth produced by enriched environments appears not to be permanent. Further, there is no window—critical period—for enhancing neural size or synaptic number in animals exposed to an enriched environment. The more general and important feature of the developing brain, then, is an initial overproduction of neurons, neuronal processes, and probably synapses. These are gradually pruned during brain maturation, initially by intrinsic mechanisms but then by extrinsic, experience-based factors, and this sculpting of the brain has a time dependence: It occurs readily in the young animal, but not nearly as much in the adult.

3

DEVELOPING BEHAVIORS

In the last chapter I described the maturation of the brain and how experience molds it, primarily by pruning. Dendritic and axonal fields of neurons are refined, synapses rearranged, and neurons even lost. There are critical periods for much of this plasticity, sensitive times early in life when various aspects of brain structure and function are particularly susceptible to alterations. In this chapter I explore examples of higher brain function development, and many of the notions described in the previous chapter hold here as well. I also discuss some new facets of brain development.

Language

Language is certainly one of the critical features that most distinguishes humans from animals. There are those who believe that the ability to speak, read, and write—and thus to communicate ideas and images that can evoke within us sensations and feelings—is what initiated our rich inner mental lives that we talk of as awareness or consciousness.

When did language first develop in humans? Human skulls of 150,000 years ago are similar in size to our own, suggesting that these early ancestors had brains like ours and were capable of language. However, there is no evidence for human behaviors that we believe link to language—rituals and complex social interactions, conceptualization and planning, art and symbolic representation—until 40,000 years ago. Thus, there is a gap of about 100,000 years during which we know virtually nothing of what was going on. Some evidence of modern human behaviors prior to 40,000 years ago has been uncovered—burials, trade, and tool making—but most paleontologists believe that it was not until 40,000 years ago that humans were fully modern and that language was universal.

All humans possess language, and as Steven Pinker remarked in his book, *The Language Instinct*, whereas "there are Stone Age societies, . . . there is no such thing as a Stone Age language." All human languages are sophisticated and complex. There are some primitive people who do not use writing, but all use complex language. (Indeed, writing is a rather new invention among all peoples. The earliest written records date back only about 6,000 years.) It is also true that the ability to speak is not essential for language; sign languages used by deaf communities can be as sophisticated as spoken languages.

Serious attempts have been made to teach language to certain nonhuman primates, especially chimpanzees. Chimps in the wild can make about 36 different sounds, almost as many as English speakers, 52. Each chimp sound typically conveys something different, whereas each sound we make (called a phoneme) usually means nothing. We string phonemes together to make words, and an educated English-speaking adult has a vocabulary of about 80,000 words.

Is it a difference in vocal tracts and speech abilities that prevents chimps and other nonhuman primates from forming words as we do? One way to test this is to teach chimpanzees sign language and this has been done, particularly by Duane Rumbaugh and Sue Savage-Rumbaugh at the Yerkes Regional Primate Center in Atlanta. They were able to teach young chimpanzees a vocabulary of about 150 words, but then the animals went no further. These chimps can communicate at about the level of a two-and-a-half-year-old child. However, this is the point at which a child's language abilities are beginning to explode. By age 3, a child typically has a vocabulary of 1,000 words and by age 4 it might be 4,000 words. Thus, humans are quite distinct from all other animals in their language capability.

Language Areas

Language is controlled mainly by areas in the cerebral cortex, and two areas have been identified as being especially important: Broca's area and Wernicke's area. However, language also depends on our ability to discriminate speech sounds, as well as to make complex speech sounds. Thus, both auditory and motor systems contribute to speech and language, and other neural systems are certainly involved too.

One of the two cortical areas especially important in language, Broca's area, is concerned mainly with the articulation and the production of speech. It is localized in the frontal lobes of the cortex near the region critical for the initiation of movements—the so-called primary motor cortex, as shown in Figure 3-1.

Broca's area is named for Pierre Paul Broca a nineteenth-century French neurologist and anthropologist, who studied people who had lost the ability to speak, a condition known as aphasia. He discovered that many of his patients had damage in that part of the cortex that now bears his name. These patients knew what they wanted to say, but their ability to articulate words was impaired. They often could not form proper speech sounds. The first patient Broca studied was called Tan because all he could utter was "Tan, tan, tan" (with an occasional oath thrown in).

Lesions in Broca's area also lead to writing deficits and even

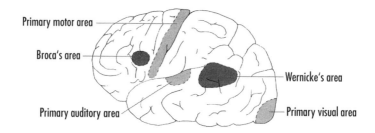

FIGURE 3-1 Lateral view of the surface of the left side of the cortex. Broca's area is adjacent to the region of the primary motor area concerned with face, tongue, and jaw movements. Wernicke's area is between the primary auditory and visual areas.

deficits in sign language, so it is clearly involved in more aspects of language than speech articulation. For example, there is general agreement that Broca's area plays an important role in grammatical processing.

The second language area is called Wernicke's area, named after Carl Wernicke, a German psychiatrist. It is found in the temporal lobe of the cortex, between the so-called primary auditory and visual areas, where sounds and visual stimuli are first processed in the cortex as shown in Figure 3-1. Patients with lesions in this area typically have difficulty with speech comprehension and with reading and writing. They can articulate words perfectly well, but their word choice is inappropriate. The words they utter are clear, but their sentences usually make no sense.

As I noted earlier, many neural systems are involved in language, so lesions in other parts of the brain can also cause language deficits. However, Broca's and Wernicke's areas are clearly key for producing meaningful language. Curiously, Broca's and Wernicke's areas are found on just one side of the brain, usually in the left cortical hemisphere (95 percent of the time), whereas most other cortical areas have representation in both cortical hemispheres. Thus, if there is damage to the left hemisphere, a patient might be totally aphasic even though the right hemisphere is completely intact. Substantial damage to the right hemisphere does not usually compromise language ability. On the other hand, if the left hemisphere is damaged early in life, many children are able to use the intact right hemisphere to learn language (see Chapter 5).

Learning Language

Linguists estimate that there are 6,000 languages spoken around the world today and thousands more were spoken at one time and are now lost. How can the human brain accommodate so many languages with so much variation? Noam Chomsky, the Massachusetts Institute of Technology linguist, studied various languages and noted that there are striking similarities among all of them. He proposed that all languages, present and past, have common grammatical principles. For example, all languages use subjects, verbs, and direct objects. The order in which these elements are positioned in sentences differs among languages, but all languages have these three classes of words. Thus, he suggested that the developing brain possesses innate neural circuits that allow for the acquisition of any of the thousands of languages now or previously spoken.

Certainly all languages share many characteristics, as Chomsky suggested, but whether the brain has innate circuitry to deal with all languages as he proposes or whether it develops at least partially as a result of experience is not certain and is a matter of much debate. It is almost certain that both innate mechanisms—nature—and learned experiences—nurture—are involved in language acquisition, although the extent to which, and how, each contributes is not settled. Clearly, innate neural circuitry must place constraints on the ability to make and perceive language, but learning is critical too, as we shall see.

Some of the most compelling evidence that there are innate mechanisms underlying aspects of language—in support of Chomsky's basic idea—is the phenomenon of creolization that occurred in seventeenth-century America. Slave owners brought together people from different African tribes that spoke quite different languages. The slaves quickly created a simplified pidgin language, based usually on the plantation owner's language. Pidgin had a crude word order but lacked a clear grammar. The children of the slaves heard only pidgin, but did not adopt it. Rather, they typically created their own languages—Creole languages—that had a grammatical structure similar to that of all other human languages.

Another bit of evidence comes from the discovery of a gene defect in a large multigeneration family that has an inherited speech and language disorder. The affected family members have problems with articulating speech sounds, identifying speech sounds, understanding sentences, and with grammatical and other language skills. The gene is inherited dominantly so that about half of the offspring of affected family members have inherited the defect—14 out of 27 offspring so far studied. The gene, called FOXP-2, probably codes for a transcription factor, a protein that interacts directly with DNA, turning genes on or off. In support of this idea, the protein contains a specific region that is known to bind a target region of DNA. Exactly what the gene does is not yet known, but an obvious suggestion is that it has a role in the development of brain circuitry related to language and speech. The gene appears to have arisen about 200,000 years ago, approximately the time that human brains assumed their modern size and when, it is thought, humans were first capable of language.

Children appear to learn language in much the same way all over the world. By one year of age, children begin to speak a few recognizable words. By 18 months, they begin to combine words and by three years, they can engage in conversation and are speaking in the language or languages to which they have been exposed. Learning a language requires no formal instruction, although hearing it spoken is critical. Indeed, it is thought that even hearing language *in utero* is involved, in that at birth infants prefer the language spoken by their mothers as distinct from other languages. And clearly, exposure to language early in life accelerates language acquisition and is essential for language development.

On the motor side of language acquisition, babies begin to

babble before 18 months and this also is critical for the development of language. We have all heard babies babbling "dadada" or "bababa," and this is the beginning of speech production by infants.

Clearly, young children acquire language much more readily than do adults, and thus it is generally agreed that there are early critical or sensitive periods for language acquisition from about 12 months to six years. Children who are exposed to a language in the first six years quickly learn to speak that language perfectly without any detectable accent. After six years it is more difficult to learn a new language, and by puberty the ability to learn a new language is dramatically reduced.

Learning a new language at 40 is similar to learning one at 20, although some people are much more adept at learning new languages than others. Linguists say that the accent for a language learned as an adult is never perfect, and that a language expert can always tell if someone has learned a language as an adult. Even many children who learn a new language between the age of 6 and puberty retain accents characteristic of their native language. The example often cited is Henry Kissinger, former Secretary of State, who came to the United States when he was eight years old. He has a distinct accent. His brother came at age 6 and is reported to have no accent.

Why do we lose the ability to speak a new language perfectly as we grow up? Youngsters are sensitive to a broad range of sounds, but they lose the ability to distinguish or make certain sounds unless they hear or produce them during the first six years or even earlier. For example, adult Japanese people cannot distinguish an "r" from an "l" sound, yet seven-month-old Japanese children can distinguish these sounds as readily as American children. By 10 months of age, native Japanese infants have already lost some of their ability to discriminate "r" from "l" sounds. American babies, on the other hand, are better at discriminating these sounds at 10 months than they were three months earlier.

The conclusion from these studies is that the period from six to twelve months is already critical for babies to learn to discriminate all different language sounds. In all languages 869 sounds or phonemes have been identified and infants six to eight months old can discriminate all of them. After that they use just a subset—those that they hear and thus distinguish. Conversely, young children can imitate virtually any sound an adult makes, but this ability is also lost with age. By one-and-a-half years, babies start to make sounds characteristic of the languages to which they are exposed, and their ability to make sounds characteristic of other languages slowly disappears.

What happens if a child is not exposed to any language for the first six to 10 years? Fortunately, there are few recorded cases, but the results are remarkably similar. The most recent example, in the 1960s, is of a young girl, Genie, who at the age of 20 months was tied up and locked in a darkened room by her psychotic father. The father and her intimidated brother only growled or barked at her for more than 10 years. When she was $13^1/_2$, she was discovered and found to be quite mute. Intensive attempts were undertaken to teach her language, but after three years of training, she was still unable to speak well; she had the language competence of a four-year-old at most. The speech she produced was labored and inarticulate. She often was unable to grasp the meaning of speech without contextual clues or gestures, and she was clearly retarded in terms of normal linguistic and comprehensive abilities. Confounding her situation was the fact that she was almost completely isolated during her imprisonment—from both sensory and emotional events—and there was some question as to whether she was mentally retarded.

However, Genie's failure to learn language was similar to that of Victor, the "Wild Child of Averyron" who lived alone in the woods in the early part of the nineteenth century. It is conjectured he was abandoned as a young child but managed to survive until he was captured at the age of 12 or 13. Victor, like Genie, never developed normal language skills, despite heroic efforts to teach them to him. There are also cases in the literature of people deaf from very early days having their hearing restored as adults who do not learn to speak effectively.

What is going on in the brain's language areas during the critical period? We cannot, of course, record from neurons in these areas as we can for the visual areas of animals during the critical periods and so we can only conjecture. It is tempting to suggest, however, that, as in the visual cortex during its critical periods, neurons can gain or lose territory, synapses rearrange and new ones form, depending on language experience. This notion might be extended to suggest that by the age of six months to a year, neural circuits have formed to discriminate and make all possible language sounds and to acquire grammar. If the circuitry is not used, it is rearranged to accommodate the native language(s) or perhaps even lost. The adage "Use it or lose it" might fit for language development as it did for visual development.

Just as with the visual system, different attributes of language acquisition appear to have their own critical periods. The critical period for making sound discriminations might be the earliest; up to six or seven months infants can discriminate all possible human speech sounds, but by 10 to 12 months, this ability is already compromised somewhat and infants might begin to show deficits. With regard to sound production, it appears that up to about five or six years, children can learn to speak a language perfectly, without an accent, although, again, some investigators believe that this critical period extends to puberty, at least for some children. The point to emphasize is that critical periods in language acquisition don't slam shut at a specific age, but there is a gradual decline in various language acquisition abilities over time, superimposed on a considerable variability among people.

With regard to grammar acquisition, a careful study of Korean and Chinese children who came to the United States showed that after 10 years of experience with English as a second language, those who arrived before age seven had a mastery of English grammar equivalent to that of native English speakers, whereas those who arrived later had grammatical skills that related to their time of arrival in the United States. Of the latter group, those who arrived earlier were more proficient than later arrivals. The grammatical skills of people who arrived in the United States after age

17 were never equivalent to those of earlier arrivals and it made little difference at what age they arrived. Thus, for grammar the critical period can extend to puberty, but it begins to close as early as seven years of age, at least for some children.

With regard to learning vocabulary, there appears to be no critical period. We can learn new words, names, and expressions throughout our lives. Certainly, children learn new words faster than adults, and for many people, vocabulary learning levels off in high school, but college students show a considerable increase in vocabulary learning as do graduate and professional school students as they are introduced to the vocabularies of new fields and areas of study. We shall return to the issue of memory and learning in adults in Chapter 4.

Birdsong

Because language is unique to humans, its development is difficult—indeed impossible—to study neurobiologically as can be done with the visual system by studying visually inexperienced or visually deprived animals, as discussed in Chapter 2. However, some systems in animals have certain similarities to human language and these systems can be analyzed in detail. Birdsong is one such example.

Of the 8,500 species of living birds, about half are songbirds. Birdsong is used for a number of purposes, including attracting mates, defending territories, or simply indicating a bird's location or presence. Birds of the same species have similar songs, but the songs can vary quite significantly over relatively short distances. Figure 3-2B shows, for example, sound spectrograms of white-crowned sparrows recorded in two locations around San Francisco Bay, Berkeley, and Sunset Beach.

The spectrograms for birds in each of these areas are similar, but surprisingly varied in the two areas. Thus, as with human speech dialects, birdsong from different geographical areas varies.

Birdsong, like human language, has great diversity. Some species, like white-crowned sparrows, have one basic song that shows

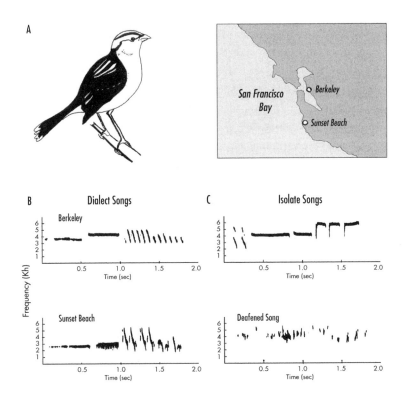

FIGURE 3-2 A: *White-crowned sparrows look identical in Berkeley and Sunset Beach, two reasonably close locations in the San Francisco area of California.*
B: *The songs of white-crowned sparrows in the two locations are distinct, however. These representations of the songs are called sonograms; sound frequency (pitch) is plotted as a function of time.*
C: *Birds raised in isolation sing, but an isolate song, simpler and distinct from normal songs. Birds deafened before they learn to sing also show very abnormal songs.*

geographic diversity, but other species sing many different songs. Certain wrens, for example, might have as many as 150 songs. The number of song units (syllables) varies considerably with birdsong, from 30 for the canary to roughly 2,000 for a brown thrasher. Again, this is comparable to the variety of speech sounds (phonemes) found in different human languages—which range from as few as 15 to more than 140. Another characteristic of song birds is that some, zebra finches, for example, sing exactly the same song throughout their lives, whereas others, canaries, for example, vary their songs; they incorporate new syllables into their songs from year to year.

How does birdsong develop? Again, we see striking similarities with human language acquisition. Young birds typically learn the songs they hear from their parents. They show a strong preference for songs of their own species, but if exposed only to songs of another species, they can acquire those songs. Indeed, young birds can develop more elaborate songs than their own species sing—if they are exposed to such songs appropriately. That young birds have a strong preference for their own species' songs suggests that they have an innate neural circuitry for those songs; however, like humans, they appear capable of acquiring certain other songs as well, so they must also have a circuitry template appropriate to accomplish this.

When first learning to sing, birds often exhibit a subsong, noises that might be comparable to babbling in human babies. The young birds next typically produce sounds that contain recognizable bits of the adult song, and finally, they begin to sing the adult song.

Song learning involves two components, song memorization and song vocalization, and a critical or sensitive period clearly exists for song memorization and perhaps for song vocalization as well. The song memorization period begins when the birds are about two weeks old and lasts for about eight weeks in white-crowned sparrows and in another model species, the zebra finch. Although birds can hear before they are two weeks old, they do not memorize their species' song if it is presented to them before the second week. Conversely, if they do not hear any song until they are three to four months old, they never sing a normal song. Interestingly, birds raised in acoustic isolation throughout the critical period do eventually sing, but only an "isolate" song that lacks both the spectral and temporal qualities of the normal song, as shown in Figure 3-2C. On the other hand, if a young bird is exposed to the normal song for only a few days during the critical period, it immediately acquires the song and sings it accurately as an adult. Young birds can learn the songs of other species, as noted above, but if the alien song differs substantially from their normal song, the birds develop isolate song singing. Furthermore, whereas birds can learn alien songs if exposed only to them, they take much longer to do this than to learn their own species' song.

If, during the beginning of the critical period for song memorization, a baby zebra finch is exposed to its normal song for one week—or long enough for the bird to memorize the song—subsequent isolation from the normal song or exposure to alien songs makes little difference. The critical period for song learning is terminated in essence after the bird has been exposed to the normal song for a week. On the other hand, if the bird is kept in acoustic isolation for several weeks after the opening of the critical period and then hears the normal song, it learns it rapidly, but this capacity clearly declines with age and is lost by three to four months. How the song is presented can also be important. Whereas presenting the normal song with loudspeakers works fine with young song sparrows, it does not with song sparrows older than 50 days old. The older birds need a live tutor bird from which to learn the song. In other species, however, a live tutor is not necessary to train an older bird—a recorded tutor works fine.

Learning to sing by birds—that is, song vocalization—is a distinct process from song memorization, and song vocalization in white-crowned sparrows occurs several months after song memorization. Song vocalization might also be constrained by a critical period, although this isn't entirely certain. Learning to vocalize clearly requires auditory feedback, so if a bird is deafened by destroying its inner ear after song memorization but before it begins to sing, it does not develop a normal song as shown in the spectrogram in Figure 3-2C. If a bird is deafened after it has learned to sing, it usually continues to sing a normal song; acoustic feedback is no longer needed.

Most birdsongs are sung only by males, and song vocalization is affected by the male hormone, testosterone. Exposing a juvenile bird to high levels of testosterone causes it to develop its song prematurely, but in an abnormal way. On the other hand, birds that are castrated before they learn to sing never develop normal singing patterns. It appears, therefore, that hormonal levels affect song vocalization in birds, and perhaps hormonal levels regulate the critical period for vocal learning. The period for vocal learning may close as the birds reach sexual maturity, at the time that testosterone and other steroid hormone levels rise in the animals.

Neural Control of Birdsong

Specific areas have been identified in the forebrain of birds that control song production and the learning of song vocalization. (The bird forebrain is analogous to the mammalian cortex.) There might also be areas specialized for song memorization, but these haven't yet been identified. The song production and vocal learning areas were first identified because of the increased size of certain nuclei (groups of neurons) in male brains. Two distinct systems have been identified—one in the posterior forebrain that is responsible for song production and the other in the anterior forebrain that is key for vocal learning. Figures 3-3A and 3-3B illustrate these two systems.

Two posterior forebrain nuclei are involved in song production, the higher vocal center (HVC) and the robust nucleus of the arcopallium (RA). When a songbird begins to sing, a wave of neural activity spreads from the HVC to the RA and then to a nucleus in the hindbrain (the hypoglossal nucleus) that contains the motor neurons controlling the vocal muscles. Lesions of the HVA or the RA make birds incapable of producing songs.

The anterior forebrain pathway also consists of two nuclei, area X and the lateral portion of the magnocellular nucleus of the anterior nidopallium (LMAN) as well as the dorsolateral thalamic nucleus called the DLM. If a lesion is made in the LMAN while a bird is learning to sing, the bird goes no further in song development but is frozen at the level already reached and is incapable of developing a mature song. Lesions in area X also prevent birds

A Song Production Pathways

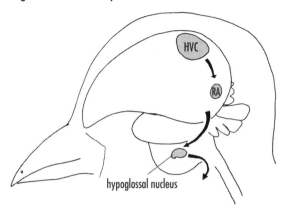

B Song Learning Pathways

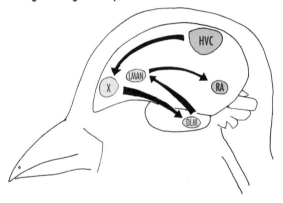

FIGURE 3-3 A: Neural pathways for birdsong production. When a bird sings, neural activity goes from the higher vocal center (HVC) to the robust nucleus of the arcopallium (RA) and then to the hypoglossal nucleus that contains neurons controlling vocal muscles.
B: When learning to sing, the HVC activates area X, which innervates the dorsolateral nucleus in the thalamus called the DLM. Axons from the DLM project to the magnocellular nucleus of the anterior nidopallium (LMAN) which, in turn, sends axons to the RA.

from acquiring a stable adult song. Circuitry studies indicate that HVC innervates area X which in turn innervates the DLM. DLM neurons project to the LMAN and LMAN axons innervate the RA, but exactly how the system works is not clear.

Neurons in both area X and LMAN respond vigorously to the sound of the bird's own song, so the notion has been advanced that the anterior pathway plays an instructive role for vocal learning. The idea is that the anterior nuclei compare the song being produced by the bird with the previously memorized song, and thus these neurons influence the circuitry in the posterior nuclei to achieve accurate sound production.

As birds reach sexual maturity and the period of vocal learning closes, neurons in all four of the forebrain nuclei involved in birdsong have been shown to bind and accumulate testosterone. As this happens, the density of dendritic spines on LMAN neurons decreases and the size of the LMAN nucleus regresses substantially. It may be that the instructive role of the LMAN is lost when sexual maturity is reached, triggered perhaps by the increase in testosterone and other steroid hormones, although this point remains controversial.

Other aspects of the HVC and RA nuclei are worth noting. For example, these nuclei are much larger in males than in females (up to five times larger in male canaries than in females). However, injection of testosterone into female canaries significantly increases the size of the nuclei. Finally, nuclei size in males appears to relate to singing skill—the larger the nuclei, the better the singer.

Now that specific groups of cells involved in sound production and vocal learning for birdsong are identified, it will be possible to work out the circuitry of the sound production and vocal learning areas and to uncover how the circuitry is being modified during development. Already some observations are pertinent. For example, axons from the LMAN are the first to innervate the RA, and only as the birds begin to sing (in this case, zebra finches) do axons from the HVC enter the RA and make synapses. A guess is that the early connections from the LMAN to RA shape the HVC to RA connections, but how this might come about is unclear.

Another direction researchers are taking is to record from individual neurons in various nuclei involved in song production and vocal learning. So, for example, neurons in HVC that innervate RA will respond to the bird's own song but not to other birdsongs. Indeed, the neurons will not respond if the song is played in reverse. Individual neurons are strongly activated by specific features of a song, but those features must occur in the context of the entire song.

Sound Localization in Owls

Studies of birdsong tell us much about song memorization and vocal learning and something of sound production. Another model system, sound localization in owls, is also highly modifiable during development and has provided further insights into issues of critical periods and how neurons and circuitry might be modified by experience.

Owls are exceptionally good at localizing sounds in space. They do this in two ways: by comparing the activation times of their two ears by a sound—intraaural timing differences—and by evaluating sound intensity differences impinging on each ear. They then construct a map of auditory space in the brain that is aligned with a map of visual space. This system enables the owl to orient its head and eyes toward an auditory stimulus. Unlike us, owls cannot move their eyes readily, so precise head positioning is crucial for auditory and visual space alignment. The auditory and visual information is integrated in the tectum, the midbrain structure that mainly receives input from the retinal ganglion cells in nonmammalian species as described in Chapter 2. When researchers record the activity of neurons in the owl tectum, certain neurons are seen to be bimodal; that is, they respond to both visual and sound stimuli. Furthermore, the neurons respond most vigorously when the sound and visual stimuli are coming from the same place. In other words, the auditory and visual receptive fields are superimposed in these neurons.

In early experiments, Eric Knudsen and his colleagues at Stanford University put earplugs in owls that reduced the intensity of sound reaching one ear. Initially, all animals made large localization errors. Adult animals never adjusted to the earplugs, but owls under the age of eight weeks at the time of plugging did. The young animals slowly compensated for the ear plugging until once again the visual and auditory maps corresponded and the owls modified how far they moved their heads in response to the auditory stimulus and could once again visualize where the sound was coming from. Interestingly, but perhaps not surprisingly, if owls with one ear plugged were deprived of vision, the compensation did not occur, indicating an instructive role for vision in the compensation process.

The converse experiments have also been done. If prisms that shift vision by 10°-20° to the right or left are put on young owls, the birds initially orient to where the sound tells them the sound source should be, but, of course, they cannot see the sound source because of the prisms. Recordings of neurons in the tectum of these owls show that the auditory and visual receptive fields no longer overlap. The visual receptive field is out of alignment with the auditory receptive field. In juvenile birds in which the prisms are kept on for six to eight weeks, the auditory receptive fields are gradually realigned with the visual receptive field. The animal does this by altering its head rotation to the sound, compensating for the prisms. Again, adult animals do not usually (but see Chapter 7) show this compensation, which extends to sexual maturity, or about 200-250 days in owls. Thus, the critical period for this compensation, like that for vocal learning in songbirds, is tied to sexual maturity.

Other observations have extended our understanding of the mechanisms underlying the establishment of the auditory and visual maps and their compensation. For example, if prisms are kept on young owls until they reach adulthood and are then removed, the animals recompensate; they completely recover over a period of weeks. This suggests that the original brain circuitry that results in the normal overlap of the auditory and visual receptive fields in the tectal neurons has persisted, even though it was not used and presumably was silent for all of this time. It is possible that new circuitry formed in the adult animal to account for this recovery, but this seems unlikely because of the experiments described next.

These experiments involved putting prisms on young owls until they compensated, and then removing them when the animals were still in the critical period. The animals and neurons recompensate, of course, but then the prisms are put on again when the owls are adults. Ordinarily adult owls do not compensate for visual distortion by prisms, but these owls do in a specific way. They compensate as adults for the same prisms they wore as juveniles, but not for other prisms—that is, for prisms that alter vision in the other direction or to a greater extent than the originals did.

These experiments indicate that synapses formed during the initial compensation period in the young bird persisted into adulthood and became active again. The discovery that circuitry modifications in young birds, and then not used for some time, can be reactivated in adults is of enormous interest and might have implications for teaching human youngsters. That is, circuitry established during early years, but then not used for years, might be more persistent than we generally appreciate. A personal anecdote might be pertinent here. When growing up, I played golf a lot, but then I essentially gave up the sport for about 30 years. When I took it up again, I found after some practice that I could play as well as I did 30 years earlier. Going beyond that level of play, however, has proved difficult.

What can we say about how the auditory maps realign with the visual maps following prism experience? The horizontal component of an auditory space map is constructed from timing differences in ear activation (interaural timing differences) in a midbrain nucleus, called the external nucleus of the inferior colliculus (ICX) as shown in Figure 3-4A.

The ICX receives its input from another midbrain nucleus, the ICC, whose neurons respond to different intraaural timing differences and which project in a topographic way to the ICX. The ICX projects to the optic tectum where it integrates the audi-

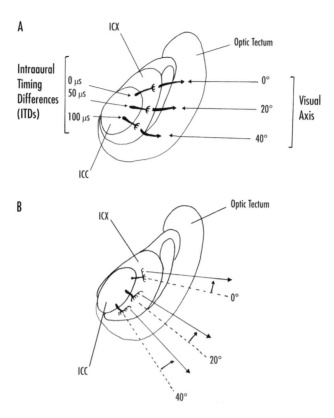

FIGURE 3-4 A: *Neural pathways involved in the alignment of auditory and visual space maps in owls. Auditory information from the central nucleus of the inferior colliculus (ICC) projects to the ICX where a map of auditory space is constructed based on timing differences of sounds reaching the two ears (ITDs). Sounds directly in front of an owl reach the two ears at the same time; therefore the ITD is 0 μs. Sounds from one side or the other have different ITDs depending on how far to the right or left the sound originates. From the ICX, signals are projected to the optic tectum where the auditory and visual maps are merged. The ICX neurons activated when the ITD is 0 μs project to optic tectum neurons that receive input from visual stimuli directly in front of the animal (0° visual axis). The ICX neurons activated in response to sounds coming from the right or left project to those parts of the optic tectum that receive visual information from the right or left in the visual field.*

B: Anatomical changes that take place in the ICX following prism compensation in the young owl. Neurons projecting into the ICX from the ICC show significantly expanded terminal arborizations, and this anatomical change could account for the realignment of the auditory and visual maps as suggested schematically in the diagram.

tory and visual information. That is, for sounds immediately in front of the animal, there are no timing differences for activation of the two ears and thus this area of the ICX projects to that part of the optic tectum receiving input from neurons on the visual axis. Moving along the tectum away from the visual axis, by 20° for example, the ICX axons projecting to that part of the tectum have the appropriate intraaural timing differences, as shown in Figure 3-4A.

Anatomical examination of the projections of the ICC neurons to the ICX show that the normal topographic projection is established early in development before prism experience exerts effects—before the critical period begins. Prism experience induces neurons to expand their terminal arborizations as shown in Figure 3-4B, so that they now synapse in regions of the ICX where they didn't synapse before. It is this elaboration of axonal terminals that appears to account for the shift in the ICX space map, which is now congruent with the optic tectum visual map.

Both observations described above, that the maps realign following removal of the prisms in an adult animal and that adjustment of the map can occur in the adult if induced in the young animal, suggest that the neuronal changes induced by prism experience are additive, not subtractive, and that perhaps both new and old synapses persist, although either might be silenced for long periods. We shall come back to this idea of synapse silencing in Chapter 4.

Imprinting—Parent Recognition

The last example I shall describe, imprinting, is quite far removed from language, birdsong, or sound localization yet it has many of the same characteristics during development. Learning to recognize one's parents occurs very early in the lives of many animals, especially birds and mammals, and has obvious positive consequences. This process is known as filial imprinting and can involve the recognition and learning of visual, auditory, olfactory, and even gustatory cues by the infant. The learning of these cues occurs during short and defined critical periods in early postnatal life.

The classic work in this area was carried out by the ethologist Konrad Lorenz in Germany in the 1930s and 1940s. He worked mainly with birds and showed that if he alone raised geese from the time of hatching, the goslings imprinted onto him and thereafter followed him around as if he were their mother.

Subsequent work by Lorenz and others showed that the critical period for imprinting is short, lasting only a few hours. In one experiment, ducks were exposed once, for only 10 minutes, to one of several male duck models that quacked. The extent of imprinting by this one exposure was evaluated five to 70 hours later by offering the ducklings a choice between the male model they had seen and a model of a female duck that also quacked. The female duck model was placed closer to the ducklings than the male model (1 foot versus 6 feet), yet a large percentage of the ducklings followed the male rather than the female, which was closer and louder.

Other experiments have extended the observations on imprinting, and again similarities were found with other types of behavioral learning. For example, if ducks are exposed to images of closely related versus non-closely related species—geese versus humans, for example, the ducks imprint on geese. Thus, as was concluded from the study of song learning in birds, there must be some innate neural circuitry that directs the learning, although it is flexible.

Imprinting on the wrong species can have long-term consequences for young birds. For example, a dove that had been imprinted on Lorenz later directed its courtship to his hand and even tried to mate with his hand when the hand was held in a certain position. Some birds imprint on inanimate objects. Chickens, for example, have been imprinted on a small bottle sitting on the back of a toy train moving around a track. But chicks, like other birds, prefer to imprint on their own species than on anything else.

An obvious next step in these experiments is to look for

changes in neuronal physiology and morphology in various parts of the brain. This has been done in a few cases, and the most intriguing results were obtained from studies on guinea fowls that had been hearing imprinted. Auditory imprinting in this species results in neurons in a region of the forebrain—the medial neostriatum and hyperstriatum (MNH)—to respond strongly and specifically to the imprinting sound stimulus. Interestingly, the dendrites of the principal neurons in the MNH of birds that were imprinted have only about half as many dendritic spines as those of nonimprinted animals. Since synapses are made mainly on the dendritic spines, these observations suggest that experience during the critical period causes a selective elimination of inputs to the MNH neurons. The loss of inputs might correlate with the decrease in the animal's ability to be imprinted with other auditory stimuli.

To conclude, although we do not as yet have much in the way of firm neurobiological evidence as to what is going on during behavioral development including critical periods, we now have evidence that a number of systems show quite similar developmental features, indicating a commonality in underlying mechanisms. Furthermore, a number of these systems appear tractable in certain animals, so it should be possible to get at the underlying neurobiological mechanisms.

PART II

THE
ADULT
BRAIN

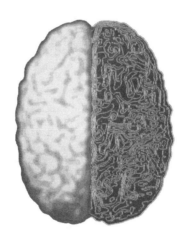

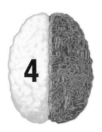

TEACHING OLDER DOGS
NEW TRICKS

The previous two chapters emphasized the point, certainly correct, that the young developing brain is much more plastic than the adult brain. Indeed, in language and birdsong development as well as sound localization in owls, puberty or sexual maturation seems to be the point when critical developmental abilities are lost. And it is common experience that motor skills—riding a bicycle or even swinging a golf club—are much more easily learned as a youngster before puberty. And once these skills are learned as a youngster they tend to stay with us for the rest of our lives.

The view that the brain becomes quite hard-wired once we become adults is a common one, but not a correct one, and recent research on the mammalian cortex has shown that it is considerably more modifiable in adults than anyone believed just a few decades ago. The beginning of this chapter focuses on these findings, then goes on to describe some of the neurobiological mechanisms that underlie cortical plasticity—in particular, what is happening in the brain when we learn and remember things.

Plasticity of the Mammalian Cortex

The notion that the adult brain is quite hard-wired goes back at least a century. Santiago Ramón y Cajal, the great Spanish neuroanatomist who many believe is the father of modern neuroscience, wrote in 1913 in the conclusion of his work on *Degeneration and Regeneration of the Nervous System*: "In adult centers the nerve paths are something fixed, ended, immutable." However, studies in several cortical areas indicate that significant modifications in cortical structure and function can occur in adults. A number of these relate to changes in response to cortical damage, but others are in response to more normal experiences. Clearly, we can learn and remember new things all our lives, and the cortex is involved in learning and memory, as we shall see. But for many decades this was thought to be a special exception, that most of the adult mammalian brain was "immutable" as Cajal suggested.

Hints that this view is not correct came first perhaps from psychological experiments, which showed that if you place ocular prisms on human beings so that the world they see is upside down, the subjects adapt within a few days and then respond to visual stimuli quite normally thereafter. When the prisms are removed, again the subjects compensate, usually very quickly (in about a day) and they again respond quite normally to visual stimuli.

This result is in stark contrast to experiments on frogs in which their optic nerves are first severed, and then the eyes rotated 180° in the head. In cold-blooded vertebrates, the optic nerve regenerates and the axons grow back to make synapses on the neurons they originally contacted. Following regeneration of the optic nerves, these animals responded exactly as if their visual world was upside down, which it was after their eyes were rotated. When feeding, they misdirected their movements by 180°: When a fly appeared in the upper right quadrant of their visual field, they reacted with a movement toward the lower left quadrant, and this aberrant behavior was permanent. The frogs never recovered from it. Thus, cold-blooded vertebrates do seem to have a much more hard-wired nervous system than mammals. Their nervous systems have other features distinct from those of mammals as well—for example, an ability to regenerate central nervous system axons. We shall return to this topic in the next chapter.

The psychological experiments using prisms on human subjects did not teach us anything about the underlying cortical mechanisms involved or even if their compensation was cortical in nature. The first evidence for structural modifications as a result of altered sensory input to the cortex came from studies carried out by Michael Merzenich and his colleagues at the University of California, San Francisco. Using monkeys, they studied how sensory input from the fingers is first processed and represented on the cortex. Somatosensory information, representing touch, pressure, temperature, and pain from all over the body surface, is first processed in the cortex along a cortical strip, called the primary somatosensory area, located just behind the primary motor area.

The surface of the body is represented on this area in an orderly and consistent way, although the body representation is not strictly proportional. This is shown in Figure 4.1, a drawing based on the studies of Wilder Penfield, a Canadian neurosurgeon who electrically stimulated the human brain during operations for epilepsy. When the primary somatosensory area was stimulated, the patients reported a sensory sensation from a specific part of the body. Those parts of the body where sensation is more acute have more nerve endings, which in turn occupy more cortical area. Thus, the face and hand take up more cortical area than other parts of the body. The same is true for the primary motor area; electrical stimuli there caused a particular part of the body to move. A greater area of the primary motor cortex is concerned with those parts of the body that we can move more precisely, such as the fingers and parts of the face like the lips, mouth, and jaw. Undoubtedly, this larger cortical representation relates to the greater dexterity and sensory acuity of the hands and face compared to other parts of the body.

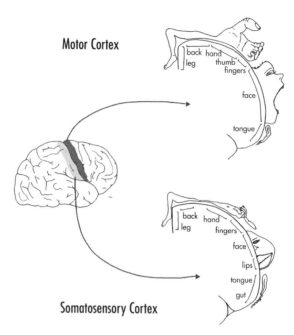

FIGURE 4-1 *The primary somatosensory and motor areas of the primate (human) cerebral cortex. The two drawings to the right show the body part associated with each area, as indicated. The body representations are not proportional; areas of the body where sensation is more acute or that exhibit finer movements (such as the hands and face) have greater cortical representation.*

These cortical representations are also termed topographic maps, and the fingers are mapped on the somatosensory cortex so that each provides sensory input to a specific region of the cortex. These regions are sequentially arranged as shown in Figure 4-2A.

By recording from individual neurons in the hand/finger region of the somatosensory cortex and determining which finger is giving a particular neuron its sensory input, Merzenich and his colleagues first found that monkeys vary substantially in how much representation their fingers have on the cortex. Some monkeys have more cortical representation for a particular finger or groups of fingers than others. But of more interest was their finding that if the sensory nerves coming from a finger are cut (called

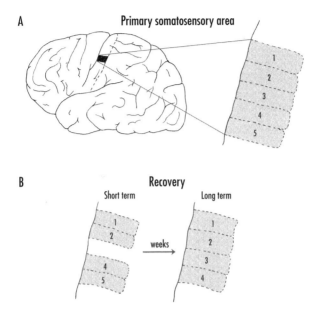

FIGURE 4-2 A: Representation of the digits on the primary somatosensory area of the monkey cortex.
B: Reorganization of the cortex following severing of the sensory nerves coming from one finger (digit 3). Initially, the area of the cortex from the deafferentiated finger was silent, but with time the area received input from neurons coming from adjacent fingers. The remaining fingers then had an increased representation on the cortex.

deafferentation), or an entire finger was removed, the representation of the fingers on the cortex changed quite dramatically. Initially, when they recorded from neurons in the area that received input from the lost or deafferentated finger, the neurons were silent as shown in Figure 4-2B. Stimulation of any finger or part of the hand produced no activation in most of the neurons. The exceptions were some neurons on the edges of the area in question, which probably shared some innervation with adjacent fingers, although this input was normally silent (a topic to which we shall return).

With time, however, it was possible to activate all the neurons in the deafferentated part of the cortex by stimulating adjacent fingers or, in some cases, other parts of the hand. This took time—weeks, even months—but the adjacent fingers gradually increased their representation and filled in the silent area. The adjacent digits now had a larger representation on the cortex than before as shown in Figure 4-2B. The conclusion from these experiments seems inescapable: New synapses and, presumably, new neuronal branches, can be formed in the adult cortex.

A question arising from these experiments is how much reorganization can take place in the adult cortex following deafferentation or loss of a part of the body. In the experiments involving the loss of a finger, the filling in of the silent cortex was relatively limited—it represented alterations in just 1-2 mm of cortex. In more extensive deafferentation experiments, carried out in monkeys by other investigators for a different purpose, the innervation to the cortex from an entire limb was cut. Eventually (the recordings were not made until 12 years after the deafferentation) the entire hand-arm region of the somatosensory cortex filled in, a distance of 10-14 mm along the cortex. Adjacent to the hand-arm region on the somatosensory cortex is innervation from the face as shown in Figure 4-1, and stimulation of the face, especially the lower jaw and chin, now activated neurons from the deafferentated area. Exactly how long it took for this reorganization of the somatosensory cortex to take place is not clear. As noted, the recordings were not made for more than a decade following the deafferentation.

Experiments by Vilayamur Ramachandran of the University of California in San Diego suggest a similar reorganization of the cortex in humans who have had a limb amputated. If the face of an arm amputee is touched lightly with a piece of cotton, the subject reports a sensation of the amputated hand being touched. Indeed, a crude representation of the hand is found on the face as shown in Figure 4-3.

Touching the cheek induces a sensation of the thumb being touched; the upper lip, stimulation of the index finger; and below the lips, touching of the little finger. It is likely that the face area has expanded into the limb area on the cortex. That the subject

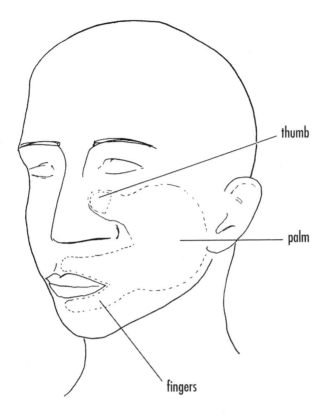

FIGURE 4-3 If the face of an individual who has lost an arm is touched lightly with cotton, the individual often reports the sensation of the missing hand being touched. The face of such individuals carries a crude sensory representation of the hand.

experiences limb sensations following light touching of the face is of enormous interest. It has been proposed that this might relate to "phantom pain," in which amputees describe sensations and even pain from amputated limbs.

Learning New Tricks

Merzenich and his colleagues also did converse experiments—to look for cortical changes following extensive stimulation of fingers. These experiments showed that if monkeys were trained to use the tips of two or three fingers to rotate a disk to get food, after several thousand disk rotations over three weeks to several months, the somatosensory cortical area for the monkeys' fingertips had expanded. Furthermore, each cortical neuron whose activity was recorded received input from a smaller area on a particular finger, suggesting a higher touch acuity for these fingertips. If rotation of the disk was limited to just one finger, the cortical expansion was limited to that finger.

Some other observations on these monkeys are worth noting. Following extensive and simultaneous stimulation of two or three fingers lasting several weeks, not only was the cortical area expanded for the stimulated fingers, but the areas had fused to some extent. An individual neuron recorded in the expanded area often had input from more than one finger, something never seen in normal animals.

Another most interesting finding was that the animals in these experiments lost the ability to move the affected fingers independently, indicating that changes had occurred in the motor system pathways as well. Indeed, a similar type of reorganization is observed in the primary motor cortex in monkeys trained to manipulate small objects with their fingers. The amount of motor cortex devoted to finger movement expands after training, whereas the cortical area devoted to wrist or lower arm movement contracts. If a monkey is trained to do a task involving the lower arm, the opposite result is observed—the digit cortical area is reduced, whereas the lower arm area is increased.

Do similar changes take place in humans who have practiced specialized motor tasks? The most compelling evidence comes from magnetic source imaging (magnetoencephalography or MEG) studies on the cortical representation of the left-hand "fingering" and right-hand "bowing" fingers of string, mainly violin, players. The cortical representation for the fingers of the left hand is greater than the cortical representation of the right-hand fingers in string players. As might be expected, the cortices of subjects who learned to play before age 12 showed more dramatic

increases in left-hand finger representation than those of musicians who began to play later in life. However, subjects who
learned to play after age 12 still had a significantly greater representation of the left-hand fingers on the cortex compared to their
right-hand fingers. Another study involving human subjects who
were proficient Braille readers and from whom MEG recordings
were made found that the recorded responses were significantly
larger for the "reading" finger compared to the same finger on the
other hand.

What about other parts of the cortex—do they show similar
plastic changes to those of the somatosensory or motor cortex?
The answer, gleaned from experiments on both the primary visual and auditory cortices of various mammals, is yes. If lesions
are made in the corresponding parts of both retinas in a monkey,
the area of primary visual cortex receiving input from that region
of the visual field is initially silent. Over a period of weeks to
months, the silent area begins to respond again to visual stimuli
placed in adjacent regions of the visual field. Eventually, the entire area can be activated by spots of light projected onto the retinas. The same type of result has been reported in guinea pig
auditory cortex. After lesions are made to the cochlea in the inner
ear that destroy the animal's ability to hear certain tones, the cortex reorganizes so that the part of the cortex once receiving input
from the lesioned area is now responsive to tone frequencies
sensed by adjacent areas of the cochlea.

Another obvious and important question is whether the entire adult mammalian brain exhibits plasticity or whether plasticity occurs only in the cortex. Recall that with visual
deprivation during the critical period in cats and monkeys, the
cortex was much more profoundly affected than was the lateral
geniculate nucleus (LGN) or retina. The same holds true for adult
plasticity. Whereas some plastic changes can be shown to occur
in subcortical structures in adults, the major site of plasticity
seems to be in the cortex. For example, Charles Gilbert and
Torsten Wiesel of the Rockefeller University showed that
whereas affected cortical areas respond again to retinal stimula-

tion two months after binocular retinal lesioning, the corresponding areas in the LGN remain silent.

A similar result was observed with monkeys trained to do a tactile task involving multiple digits in which the digits received simultaneous stimulation. Eventually a number of cortical neurons responded to stimulation of more than one digit, something that was not seen in control animals. However, recordings from thalamic neurons that receive input from the digits and that then project to the somatosensory cortex never showed multiple-digit input. This suggests that the convergence of sensory information from the different digits happens in the cortex.

What mechanisms underlie these cortical reorganizations? This is not well documented but, as noted above, is thought to reflect the sprouting of processes and the formation of new synapses within the cortex. The anatomical changes that take place in the visual cortex of rats exposed to an enriched environment, and described at the end of Chapter 2, might provide a model. As was described, most of these cortical changes could be induced in both young and adult rats. In confirmation of this idea, Gilbert's laboratory has shown that after retinal lesioning, axons from cortical neurons surrounding the deafferentated area extend collateral branches into the deprived area. Axons from the LGN, on the other hand, do not expand into the deprived area; they continue to innervate only the cortical areas they originally innervated. Thus, the reorganization of the cortex appears to represent primarily cortico-cortical plasticity, and not plasticity of input to the cortex. An expansion of cortico-cortical axons could be coupled with an increase in synapses made by terminals of new axonal branches.

Mechanisms of New Synapse Formation: The Hippocampus and Memory and Learning

There is enormous interest among neuroscientists in discovering the basis for learning and memory. The notion that learning and memory involve an alteration in synapses by, for example, increasing or decreasing synaptic strength or by the sprouting of new neuronal processes and formation of entirely new synapses goes back to the nineteenth century, but only recently has there been substantial evidence to back up these ideas. And, as neuroscientists have uncovered the neurobiological phenomena occurring in those parts of the cortex known to be involved in memory and learning, it has become increasingly clear that similar phenomena are occurring all over the brain, both during development and in the adult cortex.

The guess is that there are common mechanisms underlying the plasticity that happens during development as well as when we learn and remember new things. It is also likely that similar mechanisms come into play during the reorganization of the adult cortex that occurs as a result of peripheral lesioning or extensive training. Since these phenomena are best understood from studies on memory and learning mechanisms, I shall use this work as a model.

The Hippocampus and Consolidating Memories

A cortical area, termed the hippocampus, tucked underneath the temporal lobe of the cortex, has been implicated as a key structure in memory formation for at least half a century. Cortical areas adjacent to the hippocampus, particularly those that provide input to the hippocampus, are also critical for long-term memory formation. We have two hippocampi, one under each hemisphere of the brain as shown in Figure 4-4, and we would hardly notice the loss or degeneration of one hippocampus. However, if both are lost, along with adjacent cortical tissue, the result is devastating. We no longer remember things for more than a few minutes.

This happened dramatically in 1953 to a young man who had severe epilepsy, believed to originate in these regions of his brain, and who had his hippocampi and adjacent cortical areas removed neurosurgically. His epilepsy was cured, but he could no longer remember things for more than a short time. He retained most

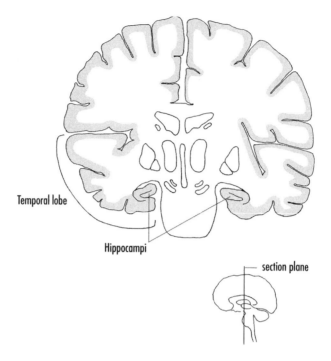

FIGURE 4-4 A vertical section through the middle of the brain (insert) showing the location of hippocampi under the temporal lobes.

memories of events that had occurred before the operation, but quickly forgot new experiences or facts. He had, in other words, lost the ability to consolidate memories.

Psychologists, particularly a Canadian, Brenda Milner, have intensively studied this patient, called by his initials HM, for more than 40 years. Virtually no changes have occurred in his ability to remember new facts or events over this period, and he remembers most of them for only 20-30 seconds, what is called short-term memory. If he keeps thinking about a new fact or event, focusing his attention on it, he can continue to recall it for longer periods (a form of short-term memory called working memory), but if he becomes distracted, he quickly forgets it. His short-term memory—the ability to remember things for seconds to minutes—appears unimpaired. What is missing is an ability to remember things long term—for more than a few minutes.

In the course of her study of HM, Milner discovered that he could learn new motor skills, which suggests that not everything we learn depends on the hippocampus. Indeed, learning new motor skills appears to involve another, noncortical, part of the brain, the cerebellum. However, what we know about mechanisms underlying cerebellar motor learning suggests that they are similar to those happening in the hippocampus.

Long-Term Potentiation in the Hippocampus

The hippocampus, like structures elsewhere in the brain, has a highly distinctive cellular organization, having three major cell types: granule cells, CA3 pyramidal cells, and CA1 pyramidal cells. Input to the hippocampus activates the granule cells, which in turn activate the CA3 pyramidal cells. These then activate the CA1 pyramidal cells whose axons provide the main output of the hippocampus. Thus, the hippocampus has a relatively simple cellular organization, shown schematically in Figure 4-5A.

In the early 1970s, Timothy Bliss and Terje Lømo, who were working in London, made a striking observation that began an explosion of research on the hippocampus and hippocampal cells that continues to this day. While recording from one of the hippocampal neurons, they found that if you provide a strong activating stimulus to the axons providing input to the hippocampus, the subsequent response of the neuron to a weak stimulus is dramatically increased. Figure 4-5B shows this schematically. The response measured is the peak voltage change that occurs in the neuron as a result of activating the synapses onto the cell with a weak stimulus. In other words, the strong or potentiating stimulus makes the synapses onto the cell more powerful; their efficacy is increased substantially. This phenomenon is called long-term potentiation or LTP.

If a strong input stimulus is repeated several times over a relatively short period, the potentiation of the synapses lasts for days

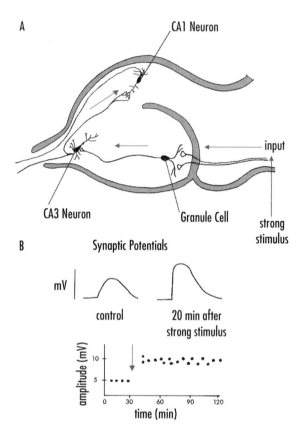

FIGURE 4-5 A: The major synaptic pathways in the hippocampus. Input axons activate granule cells, which in turn synapse on CA3 neurons. Axonal branches from the CA3 neurons innervate CA1 neurons which provide the major output of the hippocampus.
B: The generation of long-term potentiation in hippocampal cells. Synaptic potentials recorded from hippocampal neurons are increased in amplitude (potentiated) for several hours following the presentation of a strong stimulus to the input axons to the hippocampus.

or even weeks. A single strong input stimulus, on the other hand, lasts for one to three hours. Thus, investigators distinguish short-term and long-term forms of LTP and these differ to some extent in their underlying mechanisms. The major point is that with repeated strong stimuli it is possible to alter a neuron's responsiveness for long periods—days to weeks—and this, of course, immediately suggests how we remember things. That is, it suggests how neuronal excitation (an experience) can cause a long-term change in the nervous system (memory).

It turns out that not only can one induce LTP in many neurons, it is also possible to induce long-term depression or LTD. Following a strong input stimulus to some neurons, the synapses onto that neuron are decreased in effectiveness for short or long periods. Again, LTD is found in many neurons throughout the brain and can result in depressed synaptic activity for substantial periods.

Synaptic Mechanisms Underlying LTP

Weak stimuli do not induce LTP by themselves; they must be paired with a strong stimulus; then the weak stimuli show evidence of LTP as shown in Figure 4-5B. Neuroscientists talk of this as associative: Stimuli must be paired. LTP is associative in another way—to induce it requires activity in both presynaptic and postsynaptic cells, that is, both the cells making the synapses and the cells receiving synaptic input must be activated.

Why both presynaptic and postsynaptic cells must be active to elicit LTP is now understood. It depends on one of the receptor molecules present at synapses on the postsynaptic cell. A presynaptic axon terminal releases a chemical (neurotransmitter) from its synapses when it is activated. The neurotransmitter diffuses across a narrow bit of extracellular space, the synaptic cleft, to interact with receptor proteins on the postsynaptic side of the synapse. A presynaptic terminal is activated when the voltage across its membrane decreases—scientists say that the membrane is depolarized. The response elicited in the postsynaptic cell is also electrical at most synapses, but the postsynaptic response might be either an increase (hyperpolarization) or a decrease (depolarization) in membrane voltage. Depolarization of a postsynaptic cell is associated with excitation of the neuron—hyperpolarization with neuron inhibition, but this does not enter into our story of how LTP is generated.

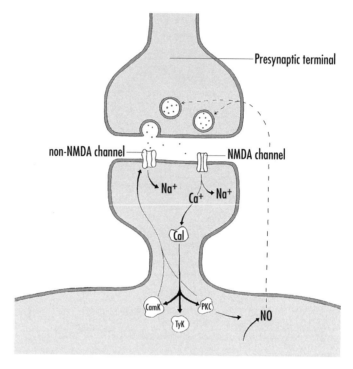

FIGURE 4-6 Mechanisms establishing LTP. When the presynaptic terminal is depolarized, glutamate is released from synaptic vesicles and interacts with channels on the postsynaptic membrane. The non-NMDA channels allow Na^+ into the postsynaptic cell, leading to depolarization of the cell and allowing the NMDA channels to admit both Na^+ and Ca^{2+} into the cell. The Ca^{2+} binds to calmodulin (Cal), which in turn activates several kinases, some of which (CamK, PKC) phosphorylate the non-NMDA channels, increasing their effectiveness in admitting Na^+ into the cell. PKC also promotes the generation of NO within the cell, which can diffuse out of the cell and into the presynaptic terminal, thereby increasing its effectiveness in releasing glutamate. Thus, LTP results from enhanced responsiveness of the non-NMDA channels to glutamate and enhanced glutamate release from the presynaptic terminal.

The neurotransmitter released at hippocampal synapses is glutamate, an amino acid. At synapses where LTP is generated, there are two types of receptor proteins in the postsynaptic membrane that interact with glutamate as shown in Figure 4-6.

Both, when activated, form channels in the membrane, allowing positively charged ions to enter the cell. Because cells are normally electrically negative inside—they have an excess of negative charges inside the cell resulting in a resting voltage or potential—the entry of the positively charged ions depolarizes the cell—the voltage across the cell's membrane decreases.

Receptor proteins that allow ions to flow across cell membranes are called channels, and the two channel types here are NMDA channels or non-NMDA channels. NMDA (n-methyl-d-aspartate) is a chemical that specifically activates the NMDA channel; it has no effect on the non-NMDA channels, and thus can be used to differentiate the two channel types.

The non-NMDA channels are like most excitatory channel proteins found at various synapses throughout the brain. When activated by glutamate, they immediately open, allowing Na^+ ions to flow into the cell and depolarize it. However, the NMDA channel works in a more complex manner, and it is key for generating LTP. If the cell is at its normal resting potential (70 mV inside negative or, conventionally, –70 mV), glutamate released from the presynaptic terminal binds to the NMDA channel, but its channel opening is blocked. It opens only if the postsynaptic cell is depolarized to some extent. The block is caused by a Mg^{2+} ion sitting in the entrance to the NMDA channel at resting membrane voltage.

Depolarization of the cell, which makes the inside of the cell more positive, pushes the positively charged Mg^{2+} ion out of the channel's mouth, and other ions can now enter it. Here, again, the NMDA channel is different; whereas most channels allow just monovalent ions that have just one charge to flow across the membrane, that is, Na^+, K^+, or Cl^-, the NMDA channel allows both monovalent Na^+ and K^+ ions and a divalent ion with two charges (Ca^{2+}) to cross the cell membrane. And it is Ca^{2+} entry into the cell that is crucial for LTP, as we shall see.

But first let's consider how the postsynaptic membrane becomes depolarized, allowing for the unblocking of the NMDA channel. This comes about by the activation of the non-NMDA channels that allow Na^+ to enter the cell and depolarize it. We can now understand why activity in both the presynaptic and postsynaptic cells is required for LTP to occur and why a strong potentiating stimulus is effective in generating LTP. Such potentiating stimuli strongly activate the presynaptic cell, causing it to release substantial amounts of glutamate. The glutamate, by activating the non-NMDA channels, depolarizes the postsynaptic neuron, thereby allowing for the unblocking and activation of the NMDA channels and the entry of Ca^{2+} into the cell.

How does Ca^{2+} influx into a neuron lead to LTP? Within the neuron, Ca^{2+} binds to a calcium binding protein called calmodulin. When activated by Ca^{2+}, calmodulin can activate a variety of kinases, our old friends that phosphorylate proteins and thereby alter their properties. Calmodulin can activate at least three different kinases in neurons, but how they increase the postsynaptic response is still not entirely clear. One possibility is that the kinases phosphorylate the non-NMDA channels, thereby increasing their sensitivity to glutamate, or alternatively by increasing the amount of Na^+ they permit into the cell following glutamate activation. Phosphorylation of non-NMDA channels is known to do this at various synapses. This, then, is a postsynaptic mechanism.

There is evidence also for an increased release of transmitter from the presynaptic terminal during LTP—a presynaptic mechanism at play. How might this come about? One suggestion for which there is supporting evidence is that kinase activation in the postsynaptic neuron results in the generation of a messenger molecule that diffuses from the postsynaptic cell to the presynaptic terminal and increases synaptic transmitter release. The gas, nitric oxide (NO), has been implicated as this messenger molecule, and the enzymes and substrates for the production of NO are present in many neurons. Figure 4-6 shows the mechanisms for establishing LTP.

Long-Term LTP

I noted earlier that there are short-term and long-term forms of LTP. A single potentiating stimulus produces LTP lasting one to three hours, whereas four or more such stimuli produce LTP lasting for days to weeks. Short-term or early LTP (E-LTP) can be explained by the mechanisms shown in Figure 4-6, but long-term or late LTP (L-LTP) involves more elaborate pathways and more permanent changes in the cells and their synapses. For example, new protein synthesis occurs in L-LTP, but not in E-LTP.

Figure 4-7 shows schematically the mechanisms involved in L-LTP. In this process too, Ca^{2+}-activated calmodulin is involved.

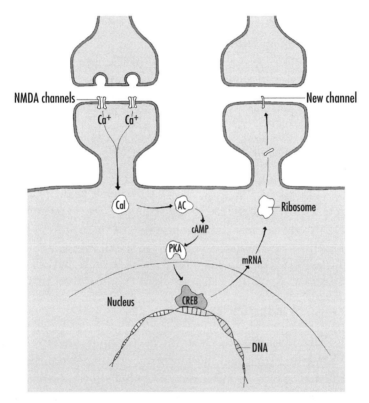

FIGURE 4-7 Mechanisms underlying L-LTP. Ca^{2+} entering the cell via NMDA channels activates calmodulin (Cal) which in turn activates the enzyme adenylate cyclase (AC). AC catalyzes the production of the second-messenger cyclic AMP (cAMP) which then activates a kinase (PKA). PKA phosphorylates a transcription factor (CREB) which interacts with DNA in the cell nucleus leading to gene expression and the production of messenger RNA (mRNA). mRNA moves out of the nucleus and interacts with ribosomes which results in the production of new protein. The newly made proteins can, for example, make new channels that are inserted into the membrane.

Scientists believe that if sufficient calmodulin is activated, it interacts with an enzyme called adenylate cyclase (AC). This enzyme converts a molecule called adenosine triphosphate (ATP) to a smaller cyclic molecule, cyclic AMP (cyclic adenosine monophosphate). Cyclic AMP is called a second-messenger molecule because it can readily diffuse in the cell and interact with proteins. For the production of L-LTP, the cAMP interacts with a specific kinase protein (PKA) and this leads to the phosphorylation of a transcription factor called CREB (cyclic AMP response element binding protein).

Transcription factors, as described in Chapter 1, interact directly with those regions of genes in the nucleus (the promoter regions) that turn on or off gene expression. When a gene is to be turned on—that is, expressed—the code for the protein to be made is transcribed from the gene's DNA into a piece of another, slightly different nucleic acid, RNA. The messenger RNA (mRNA) moves out from the nucleus of the cell to the cytoplasm where it is translated into protein by structures called ribosomes. In this way, new protein is made that can lead to the strengthening of synapses by, for example adding new channel proteins to them, to the formation of entirely new synapses, or even to the development of new branches made by the neuron.

LTP Elsewhere in the Cortex

LTP has been observed widely in the cortex, including in the primary visual, auditory, and motor cortices. It is usually easier to elicit LTP in the hippocampus than elsewhere in the cortex, probably because of the relatively simple and straightforward hippocampal cellular organization shown in Figure 4-5. Nevertheless LTP has been recorded in many brain structures. Furthermore, NMDA receptors have been identified throughout the cortex, so scientists believe that the mechanisms for eliciting LTP in the hippocampus probably apply elsewhere.

There is enormous interest in long-term LTP (L-LTP) in particular, because it provides a compelling model of how new neuronal processes and synapses can be generated in both the young and adult brains and how synaptic circuitry can be altered in the brain as a result of experience. On the other hand, we know little as yet about the natural stimuli required to elicit LTP in various parts of the cortex and elsewhere. In the laboratory, LTP is usually generated in specific neurons by applying artificial stimuli to the axons providing input to the neurons. Some recent studies, on the other hand, show that certain auditory and visual sensory stimuli can induce LTP-like potentiation in some neurons.

A feature of LTP generation of developmental interest has been observed in the visual cortex of young rats and cats. Whereas LTP can be generated in the visual cortex of young animals, the ease of its generation is age dependent. As the animals mature, it becomes harder and harder for them to generate LTP in visual cortical neurons. This decline in LTP generation roughly parallels the critical period for the solidification of the ocular dominance columns in the cortex. The idea that the two phenomena are related is strengthened by the finding that dark-rearing, which prolongs the ocular dominance critical period, also prolongs the time it is possible to generate LTP in the visual cortex.

A similar critical period for the generation of LTP has been shown for neurons in the somatosensory cortex. Thus, LTP-like mechanisms might underlie the pruning and refinement of synapses that happen in early development, as well as the rearrangement and addition of synapses that occur both during critical periods and in the adult cortex. The bottom line is that, while the underlying mechanisms of all these phenomena are likely to be similar, their extent and ease of generation might vary with age and brain region.

Neuromodulation and Silencing Synapses

There are mechanisms by which neurons and their synapses can be modified other than through NMDA receptors and LTP. This is frequently referred to as neuromodulation and is generally thought to involve changes in neurons that last from minutes to hours. Neuromodulation does not usually involve protein synthesis, although it can. Whereas it might take 15-20 minutes for LTP to be generated fully in a neuron, neuromodulatory effects typically begin within 15-20 seconds of their initiation.

Many of the same players involved in the various forms of LTP are also involved in neuromodulation. For example, the best-characterized neuromodulatory system involves the second messenger, cyclic AMP, and protein kinase A. Figure 4-8 shows a scheme for neuromodulation involving these substances.

At synapses where neuromodulation occurs there are receptor proteins in the postsynaptic membrane that interact with the substance released from the presynaptic terminal (a first messenger). In the case of neuromodulatory synapses, the chemical substances released, the first messengers, are most often monoamines, such as dopamine or serotonin, or small peptides, the neuropeptides. The membrane receptors they bind to don't form channels in the membrane, but interact with one of a set of proteins called G-proteins. The G-proteins serve to activate enzymes such as adenylate cyclase, which produces second-messenger molecules such as cyclic AMP. Most often, cyclic AMP activates the PKA kinase, which then can act at virtually any level of the cell—at the membrane to alter the properties of membrane proteins including channels, in the cytoplasm to activate or inactivate enzymes, or in the nucleus to turn genes on or off by the activation of transcription factors.

We now know a variety of such neuromodulatory pathways involving the production of different second messengers, the activation of many different kinases and so forth. One of the striking phenomena that result from the activation of these pathways is the strengthening or weakening of synaptic activity. Indeed, synapses can even be silenced in this way. That is, they are present but they are not functioning or they are functioning so weakly that they are ineffective. Neural plasticity that occurs relatively rapidly in the brain—and there are many examples—probably relates more to neuromodulatory effects than to LTP and structural alterations in the nervous system.

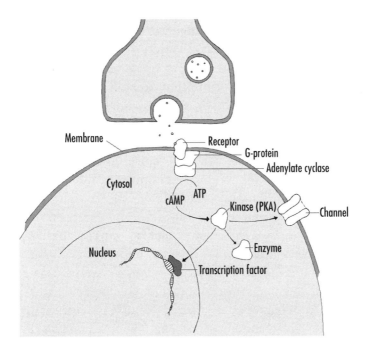

FIGURE 4-8 How neuromodulation can affect a neuron. The neuromodulatory substance released from the presynaptic terminal activates a membrane protein called a receptor. When activated, the receptor protein interacts with and activates a second protein (G-protein), which in turn leads to the activation of an enzyme, in this case adenylate cyclase, which converts ATP to cyclic AMP (cAMP). The cAMP activates the kinase PKA, which then can phosphorylate proteins in the cell membrane (channels), cytosol (enzymes), or nucleus (transcription factors)—thereby activating or inactivating them and altering the properties of the neuron.

I noted earlier that after the removal of a finger from a monkey some neurons on the edges of the cortical area receiving input from that finger quickly—within minutes—give responses to stimulation of adjacent fingers. It is likely that this activity is the result of silent or weak synapses now coming into play or being greatly strengthened as a result of activation by a neuromodulatory pathway. On the other hand, it could be the result of the simple removal of inhibitory input to the adjacent finger neurons that are providing this input. While it is not clear which is responsible, it is clear that a variety of mechanisms can alter synaptic strength and circuitry in the adult brain—from simple synaptic excitation and inhibition, to strengthening or weakening of synaptic strengths by neuromodulatory mechanisms, to neurons sprouting new branches and forming new synapses by mechanisms such as L-LTP.

These mechanisms can have quite different time courses, from milliseconds for ordinary excitatory and inhibitory synaptic effects, to minutes for neuromodulatory effects, to hours, days, or weeks for substantial remodeling of neurons and the formation of new synapses by mechanisms such as L-LTP. These different mechanisms are not always distinct, but represent a continuum of phenomena that generate an enormous richness of cortical plasticity and underlie a variety of mental phenomena.

I end the chapter by giving two examples of cortical plasticity that relate to perceptual phenomena. The first involves the modifications of the properties of orientation-specific neurons in the visual cortex. In Chapter 2, I described the receptive field properties of such neurons, shown in Figure 2-3A. Ordinarily these neurons respond best when a bar or edge of light with a specific orientation is present in the appropriate part of the visual field, that is, within the receptive field of these neurons. Bars or edges of light of any orientation do not activate these cells when projected onto the retina outside the receptive field. These neurons, then, appear wired to respond to a bar of light of a particular orientation in a restricted region of the visual field.

However, if several bars of light of a somewhat different orientation are placed outside the receptive field of the recorded cell and then the receptive field orientation of the neuron is reexamined, the neuron's orientation selectivity might be altered as shown in Figure 4-9A.

The bars of light outside the receptive field do not activate the recorded cell when they are applied; their effect is to change silently the properties of the recorded cell. Not every cortical neuron shows the effect and the change in orientation selectivity appears limited—to about 10° or so. Nevertheless, it shows that even

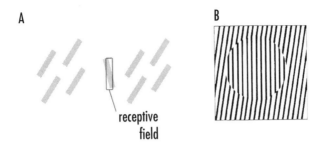

FIGURE 4-9 A: The orientation tuning of a cortical neuron can be altered by up to 10° by the presence of oriented stimuli outside the receptive field. The orientation selectivity of the neuron (clear bar) is tilted to the left (shaded bar) when bars of light tilted to the right are presented outside the receptive field.
B: The tilt illusion. The lines in the circled area appear tilted to the left although they are vertical.

fundamental receptive field properties of cortical neurons such as orientation selectivity have some plasticity.

What causes these neurons to be plastic is not entirely clear, but the answer might be found in collateral branches of cortical neurons that can extend several millimeters horizontally in the cortex. Although these collateral branches have been known anatomically for some time, their function is obscure. Indeed, it might be that the synapses they make are often silent and are activated only under specific conditions such as during the experiment just described.

What consequence might such effects have? Visual physiologists have proposed that the phenomenon described above relates to the "tilt illusion," in which the perceived orientation of a line depends on the orientation of surrounding lines, and not on the actual orientation of the lines. Figure 4-9B shows this. The lines in the middle of the figure are vertical, but because the surrounding lines tilt to the right, the central lines appear to tilt left. The more general purpose of such effects might have to do with integrating information from different parts of a complex scene—helping to explain, perhaps, the effects of context in visual perception.

Not only can peripheral stimuli outside the receptive field of

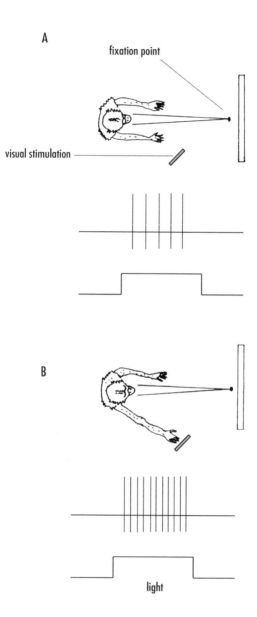

A

fixation point

visual stimulation

B

light

a cell affect a cell's responsiveness, but attention of the subject can also alter responsiveness. Robert Wurtz and Michael Goldberg at the National Institutes of Health first showed this. If the activity of a neuron along the visual pathways in a monkey is recorded, the strength of that activity can be increased by having the animal focus attention on the stimulus that drives the recorded neurons.

Figure 4-10 shows the experiment. A monkey is trained to keep its eyes on a fixation point in front of it. A visual stimulus then comes on elsewhere in the visual field, but within the receptive field of the recorded neuron. The visual stimulus activates the recorded neuron, but if the visual stimulus is modest in strength, so is the recorded response (in this case the action potential output of the cell, which consists of just four events).

If instead the monkey is trained to focus attention on the stimulus by reaching out and touching it when it appears (for a reward, of course), without removing its eyes from the fixation point, the activity of the recorded neurons to the stimulus is considerably enhanced.

We do not know the mechanisms underlying the plasticity illustrated by these two examples. Almost certainly they don't involve structural changes in the brain because they occur very quickly and are rapidly reversible. However, they show how apparent rearrangement of neural circuitry and altering neuronal responsiveness can occur in the cortex, probably as a result of increasing or decreasing the effectiveness of synapses; that is, whether the synapses transmit stronger or weaker signals or even transmit signals at all.

FIGURE 4-10 *The effect of attention on the responsiveness of a neuron in the cortex.*
A: *The neuron being recorded is activated by a visual stimulus in the periphery of the visual field. When the monkey is looking at the fixation point in the center of its visual field, the response of the neuron to the peripheral stimulus is minimal.*
B: *If the monkey attends to the peripheral stimulus by touching it, the response of the neuron is significantly enhanced, even though the monkey has not shifted its gaze from the fixation point.*

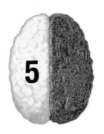

5

CONTROVERSIES: NEW NEURONS AND GENES AND BEHAVIOR

The previous chapter described various ways cortical circuitry can be modified—from simply altering the strength of synapses to neurons extending new branches and making new synapses. It deftly avoided one of the most contentious questions of the day: Can new neurons be generated in the adult mammalian brain? New neurons can be generated in nonmammalian brains, as I shall describe below, but what about the mammalian brain? This is a hotly debated issue that has implications not only for understanding the adult mammalian brain, but perhaps even more so for the aging brain. It has implications too for the extent of plasticity that can occur in the mammalian brain, also a topic not touched upon in the last chapter but one of vital importance.

The second half of the chapter considers another very contentious issue—the relation of genes and behavior. This is a topic about which reams have been written and which generates enormous heat because of its obvious social implications and the high stakes involved. If our behavior is fixed by our genes, why should we try to alter it by education or social programs? Obviously, in half a chapter I can only touch on the topic, but hard neurobiological facts are few and far between here. This discussion, then, can only point the way; its resolution—if indeed it even can be resolved in a satisfactory way—is not yet clear.

Generating New Neurons

Neurons differ from other cells in the body in at least two important ways. First, neurons have an absolute requirement for oxygen. Deprived of oxygen, mammalian neurons die in just a few minutes. Other cells can survive for some time without oxygen and even function for a while by a fermentation-like process that breaks down sugar (glucose) in the cells to smaller molecules, thereby producing the energy needed for them to keep functioning. During a 100-yard race, a runner's leg muscles probably use up the available oxygen in the first 30 yards or so. The rest of the race is run without oxygen (anaerobically). After the race, the muscle cells break down the smaller molecules using newly available oxygen. While this is going on, a runner typically breathes hard to repay the oxygen debt built up in the muscles during the race. After strenuous activity, muscles often feel sore, due in part to the buildup of the breakdown products of glucose, particularly lactic acid, during the anaerobic phase of muscle activity.

Neurons, on the other hand, cannot survive anaerobically; they stop functioning shortly after the oxygen runs out. After a heart attack or stroke in which blood flow to the entire brain or part of the brain is shut off, neurons begin to die after just four to five minutes. Other tissues can survive much longer, so that if oxygen is eventually restored, they survive, but the brain does not. Thus today, with our sophisticated life support systems, someone suffering a heart attack might be left permanently brain dead or with permanent brain damage even though other organs such as the heart, liver, and kidney recover completely.

The brain's requirement for oxygen is so acute that when a part of the brain is active, blood flow to that region rapidly increases. The rise in blood flow to an active part of the brain can be measured and is the basis for certain imaging techniques, including positron emission tomography (PET) scanning and functional magnetic resonance imaging (fMRI), that enable neuroscientists to observe which parts of the brain are especially active when a subject is carrying out a task. These noninvasive imaging techniques are adding enormously to our understanding of human brain function.

The other way neurons differ from other cells is that once they have differentiated from a neural progenitor cell into a specific type of nerve cell, they never divide again. As I pointed out in Chapter 1, the brain initially overproduces neurons, so that during the first year of life humans have as many neurons as they ever will. Thereafter, the number of neurons decreases throughout life. A controversial issue is how much cell death normally occurs throughout the life span—a topic that will be taken up in the next chapter.

Because neurons can't divide again after differentiation, and because anatomists observed that after injury to various parts of the brain no new neurons seem to appear, it has generally been believed that new neurons are not generated in the adult mammalian brain. This is in marked contrast to other parts of the body where, after injury, new cells are rapidly generated and the injury is repaired. A cut on the skin heals quickly as new cells are generated and fill in the defect. Usually within a few days no sign of the cut can be detected. "New glial cells are generated in the mammalian brain throughout life, but not neurons" was the dogma.

Some recovery, of course, is usually observed following brain injury or stroke, but this is thought to be the result of the recovery of cells damaged but not killed by the lack of oxygen, or the remaining brain cells taking over for the lost cells. For example, as has been clearly shown and was discussed in the last chapter, after brain injury, neurons sprout new branches and form new synapses, allowing for at least some recovery. The fact that neuronal brain tumors are not seen in adults is another piece of evidence supporting the notion that new neurons are not generated in the adult brain. Brain tumors in adults are mainly glial cell tumors or tumors arising from other nonneuronal cells. In children, rare neuronal cell tumors are observed, mostly derived from a pool of neuronal precursor cells that continue to proliferate for a time after birth.

A New View

The dogma that the mammalian brain cannot generate new neurons was recently challenged. The challenge arose, not from disputing the notion that neurons, once differentiated, no longer can divide, but from the discovery that two regions of the adult mammalian brain, the hippocampus and the cerebral subventricular zone (SVZ), retain neural stem cells that generate new neurons. These new data are convincing and fly in the face of the old dogma. But what are these new neurons up to? What role do they play? As I shall describe below, these questions are yet to be answered.

The finding that two brain regions can generate new neurons has spurred investigators to examine other brain areas, including the cerebral cortex, for similar neural progenitor cells. Claims that such cells exist in the cortex have been made with much fanfare, but these findings are hotly disputed. Glial cells continue to divide in the adult cortex and one view is that the dividing cortical cells are glial progenitor cells. Further confusing the picture is the view of some researchers that glial progenitor cells can and do become neurons. Others disagree, believing that new neurons are generated in the cortex, so the story is murky at the moment. Most neuroscientists are quite agnostic on the issue of whether most regions of the brain are capable of generating new neurons.

More significant, perhaps, than whether there are new neurons produced in the mammalian brain is the question of the role they play: What is their significance? Again, here the story is incomplete, but also strange. We know most about the hippocampal cells, which Fred Gage and his colleagues at the Salk Institute in La Jolla, California, have studied extensively. Less is known about the SVZ cells, which supply the olfactory bulb (and only the olfactory bulb as far as we know) with new neurons.

The peculiarity of the new hippocampal neuron's story comes from several observations. First, the new neurons generated are limited to just one region of the hippocampus—the input region where the granule cells are found (see Figure 4-5). Also, the newly generated cells are mostly quite transient—many of them disappear within a few weeks of being generated, although some might be found in the hippocampus for as long as 12 weeks in monkeys and perhaps up to two years in humans. However, they do not appear to be permanent, as are most of our neurons that last a lifetime.

Further, the number of new neurons generated decreases substantially with age. For example, in rats the number of new neurons generated at 21 months—about the midpoint of their life span—is less than 10 percent of the number generated at six months, raising the intriguing possibility that this generation of new neurons could be a slow tailing off of a developmental process.

As yet, clear-cut evidence that the newly generated neurons incorporate themselves into the hippocampal circuitry is lacking, although we assume this is the case. Neurons in the rat hippocampus are known to be sensitive to stress, and exposure to stress decreases the numbers of new neurons generated. This seems to be due to a decrease in neuronal proliferation, caused perhaps by raised levels of stress hormones, the glucocorticoids. Indeed, the treatment of nonstressed rats with corticosterone, the main rat stress hormone, decreases the proliferation of new cells in the hippocampus. On the other hand, certain hormones, including the female estrogen hormones, increase the number of new neurons generated in the hippocampus.

At present, we can come to only a few conclusions with regard to the role of the new neurons in the hippocampus, although it is clear that they are generated there. The major question, whether the generation of new neurons in the adult mammalian brain is the exception or the rule, remains unanswered.

Neurogenesis and Birdsong

The study of songbirds has provided important insights into adult brain neurogenesis. Indeed, neurogenesis in birdsong areas of the adult avian brain might be a model for the adult mammalian brain. Fernando Nottebohm and his colleagues at the Rockefeller University in New York are leading researchers in this area and I describe a number of their findings below.

Clearly, the brains of adult songbirds generate new neurons, but only in the forebrain. Furthermore, new neurons are added to just a few narrowly defined areas and to specific circuits in the forebrain, and these areas and circuits relate mainly to song production (see Figure 3-3). For example, newly generated neurons are found in the HVC nucleus, which sends axons to the RA nucleus, but no new HVC neurons have been seen sending axons to area X. We know, however, that there are neurons in the HVC nucleus that send axons to area X (see Chapter 3), indicating both a cell and circuitry specificity for the new neurons. Present evidence suggests that the newly generated neurons replace ones that have died, but only a few of the neuronal cell types present in the birdsong areas can be replaced. Indeed, only three types of neurons are replaced out of a total of about 24 types in the songbird circuitry. Two of the three are in the HVC nucleus, the third in area X.

The new neurons, it is believed, are generated strictly to replace dead ones. Once the HVC has its adult complement of neurons (at about 8 months of age), there is no increase in total numbers of HVC neurons. If HVC neurons are destroyed in the adult, there is a sharp increase in the generation of new neurons. However, only HVC neurons projecting to RA are generated; neurons projecting to area X are not; they remain depleted. It is estimated that about half of the HVC neurons projecting to RA are replaced every six months in the normal adult bird.

Like the neurons in the mammalian hippocampus, many of the newly generated neurons in the bird forebrain disappear (and presumably die) two to three weeks after they are generated. Only about one-third of the newly generated neurons remain after 30-

40 days. Just as in the developing brain, there appears to be an overproduction of newly generated neurons, but unless the new neurons can find synaptic targets, they die.

The situation with regard to the maintenance of new HVC neurons is also complex but interesting. For example, if singing is suppressed in a songbird, many of the newly added neurons disappear from the HVC. Furthermore, singing is seasonal in many birds, and during periods of increased singing, the HVC gets bigger largely as a result of the survival of more of the newly generated neurons. Increased singing and the survival of more HVC neurons are both related to increased levels of testosterone in the blood, which leads to an increase in the growth factor, BDNF. (See Chapter 1 for a discussion of BDNF.)

I note one other observation before I attempt to interpret what these observations might mean. Although the songbirds studied by Nottebohm and his colleagues can live for up to 10 years, very few of the new neurons (the ones scientists can monitor) live for as long as a year. This suggests that unlike most neurons in the brain, which live for the life span of the animal, the neurons that are replaced in the songbird brain are very labile. Thus, it appears that the generation of new neurons in the bird brain is very much the exception rather than the rule. Only a few types of neurons can be replaced and, indeed, need to be replaced routinely. Whether the same holds true for the mammalian brain is not clear, but the similarities between the bird and mammalian brain in this regard suggest that it might be the case.

On the other hand, the fact that even some neurons can be replaced is intriguing and suggests that perhaps neuroscientists can discover ways to encourage the replacement of old neurons with newly generated ones within the brain or with cell or brain tissue transplants. Could we induce the rare neuronal stem cells present in the brain to proliferate profusely and become a variety of types of neurons, or if neural stem cells are transplanted into the brain, could we induce them to differentiate into various types of neurons and incorporate themselves into the brain's circuitry? We don't know the answers.

Some brain cell transplantation experiments have been carried out in attempts to intervene in Parkinson's disease, and limited success has been reported. In Parkinson's disease, there is a depletion primarily of the cells that release the neuromodulator dopamine (see Chapter 4). The cells transplanted into the brain appear to secrete dopamine and help in that way. The transplanted cells appear to work, at least for a while, although side effects are often a problem. There is no evidence that the transplanted cells incorporate into the circuitry of the brain or replace the dopaminergic neurons that have died in Parkinson's disease. I shall discuss Parkinson's disease further in the next chapter.

Clues from Cold-Blooded Animals

Although the evidence for the generation of new neurons throughout the mammalian brain is weak and controversial, the same is not true for cold-blooded vertebrates. In many fish, for example, the retina of the eye grows throughout the life of the animal. Around the retina's periphery is a marginal zone, consisting of neural stem cells that are continually dividing and generating all of the retina's five types of neurons and its major type of glial cell as shown in Figure 5-1.

New retina is continually being formed in a ring-like fashion around the margin of the eye, much as a tree grows its trunk. Newly formed ganglion cells project into the tectum in the midbrain (see Chapter 1), and to accommodate them, new tectal neurons are also added. In adult goldfish about 50 new ganglion cells are added to the retina each day.

Whereas most of the stem cells in the fish retina are found in the marginal zone, other stem cells are scattered throughout the outer half of the retina as shown in Figure 5-1. These are called rod progenitor cells because they ordinarily give rise to new rod photoreceptor cells. As the eye grows, it is important to increase the number of rod cells across the retina to maintain rod cell density and hence light sensitivity at optimal levels. However, when the retina is damaged, the rod progenitor cells are capable of forming all the retinal neurons and repairing the damage. Even very

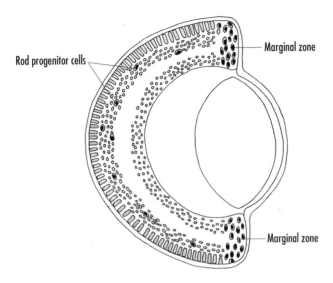

Rod progenitor cells

Marginal zone

Marginal zone

FIGURE 5-1 A sketch of a fish eye showing the marginal zone where there are stem cells that divide and generate new neural and glial cells throughout the life of the animal. There are also some stem cells, called rod progenitor cells (arrows), found throughout the outer retina. These cells ordinarily generate new rod photoreceptor cells, but if the retina is damaged, they generate all retinal cell types, thus repairing the damage.

large lesions are completely repaired and vision is restored to that part of the eye.

Why the cold-blooded nervous system can regenerate itself and the mammalian brain cannot, or can only to a very limited extent, is not understood, but it is obviously an area needing further study. It is curious that the brains of cold-blooded vertebrates are much more plastic in terms of neuronal replacement and repair than mammalian brains, but at the same time they appear much more hard-wired and perceptually less plastic than mammalian brains. Recall the example of inverting prisms: Humans rapidly adjust to an upside-down world, but when a frog's world is turned upside down, it remains that way. Is there a trade-off of one type of plasticity for the other? No one knows.

Cold-blooded vertebrate neurons differ from mammalian neurons in other ways, particularly with regard to the regeneration of axons. Whereas the axons of peripheral nervous system neurons (that is, axons outside the spinal cord and brain) in both mammalian and nonmammalian species regrow after being severed, central nervous system axons in mammals do not. Thus, spinal cord injuries that cause the severing or loss of central nervous system axons result in a permanent paralysis of the body and limbs served by those axons. Not only motor function, but also sensation is lost. Such people can never walk again. If the spinal cord injury is just below the neck, they might also lose the use of their arms and require assistance breathing, necessitating the use of a respirator. The actor Christopher Reeve, whose spinal cord was crushed in a riding accident, is an example of this type of devastating injury.

It turns out that the determinant of whether an axon regenerates resides not in the axon itself, but in the glial cells that cover and contact the axons. All axons are ensheathed by glial cells, and the glial cells form a lipid layer (the myelin sheath) around most axons to insulate them. The glial cells in the peripheral nervous system that ensheath the axons are distinct from those in the central nervous system. If severed central nervous system axons are brought into contact with peripheral nervous system glial cells, at least some of the axons regenerate, as first shown by Albert Aguayo and his colleagues in Montreal. It seems that two things are going on. The peripheral nervous system glial cells appear to release substances that promote the regeneration of axons, whereas central nervous system glial cells release a factor or factors that inhibit the regeneration of axons. Identification of these factors is currently being attempted with some success, but the story is not yet complete.

In cold-blooded vertebrates, central nervous system axons do regenerate, so this is another avenue of research to pursue—to attempt to find differences in the nerve and glial cells between cold-blooded and mammalian species. Such studies might teach us not only why axons regenerate in one situation but not another but, even more importantly, how we might induce the generation of new neurons throughout the brains of warm-blooded animals.

How Plastic Is the Mammalian Brain?

Another issue is how much reorganization can occur in the mammalian brain, either young or old. I pointed out in the last chapter that when both the two hippocampi and their associated areas are destroyed, a person can no longer consolidate memories. Such people must live in the present and can never again remember new facts or events. There is no recovery.

Adults who have a massive stroke on the left side of the brain are often left mute and show little or no improvement in speaking abilities over time. However, a child can be very different. Occasionally, and fortunately it is very occasionally, it is necessary to remove virtually a whole cortical hemisphere in a child because of a condition known as Rasmussen's encephalitis. This disease causes a debilitation of one hemisphere that results in severe disruption of the other. Surgical removal of the diseased hemisphere permits the other hemisphere to function. The question is how well these children do after surgery, especially if it is the left hemisphere with its language centers that is removed.

The answer is that they do remarkably well. They learn to speak again over months, and motor function on the right side also returns. Neither speaking itself nor motor function ever becomes entirely normal, but the amount of recovery is remarkable nevertheless. Language comprehension is better than speaking for such children and, as is expected, the younger the child is at the time of the surgery, the better and quicker is the recovery. No one older than 15 has ever regained much function.

In these children, one hemisphere takes on the work of both; a new hemisphere does not develop and fMRI scans show that the remaining hemisphere does not increase greatly in size and fill up the skull. Rather there is a large void where the removed tissue was, as shown in Figure 5-2.

Exactly what is going on in the remaining hemisphere is not known, but presumably there is extensive remodeling of the neural circuitry, growth of new branches by the neurons, and new synapse formation. This has not yet been explored.

Normal   Following surgery

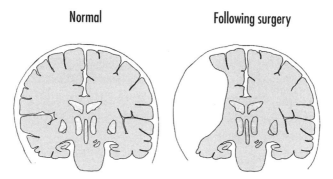

FIGURE 5-2 On the right is a sketch of the brain after surgery of a young boy who suffered from Rasmussen's encephalitis. Although much of the left hemisphere, including the language areas, was removed, the youngster regained speech and other functions normally controlled by the left hemisphere.

Is there a similar takeover of function by one part of the brain for another in all circumstances in children? The answer appears to be no. If, for example, there is a bilateral lesion of the cortex so that comparable areas of both hemispheres are damaged or destroyed, the functions mediated by those cortical areas do not recover. This has happened occasionally during operations on infants when oxygen levels in parts of the brain fell precipitously. The children were then left permanently mute and/or blind or with some other serious neurological deficit. It appears that recovery depends on a brain area in one hemisphere remaining intact.

Genes and Behavior

No one would deny that genes are critical for behavior, but the question of how and how much is a vexing one. It is almost needless to say that the relationship is an exceptionally complex one, and because of the enormous stakes involved—ranging from how we raise our children to how we deal with societal issues—the debate is hot and heavy. We read every day about a newly discovered gene that accounts for this behavior or that—from bed wetting to homosexuality to doing mathematics. We also read reports from the other side of the debate, arguing that behavior is, as the title of one book indicates, *Not in Our Genes*. The truth is somewhere in the middle, of course, and we are still a long way from sorting it all out.

One classic example often cited as showing a relation between genes and behavior is schizophrenia, a devastating cognitive disorder that results in thought and mood disorders, delusions and hallucinations, and the withdrawal from social interactions. The judgment of schizophrenics is markedly impaired and they can lose all contact with reality. It has been known for some time that there is an inheritance factor in susceptibility to schizophrenia, and the question is, how large is the inheritance factor? Studies of identical twins, whose genetics are the same, concluded that if your identical twin is schizophrenic, the chances that you will become schizophrenic are about 50 percent. On the other hand, fraternal twins have an incidence of about 15 percent and schizophrenia in the general population is only 1-2 percent.

"Ah," says one side, "see how important genetics is for this behavioral disorder!" But the other side comes back with: "What about the 50 percent that don't become schizophrenic? The genes are the same; environment is critical!" And, of course, both sides are right and so we are in a muddle. What can we say?

Huntington's Disease—A Clearer Case

An inherited neurological disorder, Huntington's disease, is a clearer example of how one gene can cause a devastating disease affecting behavior. The disease is fortunately rare—affecting about 1 in 10,000 people. It begins usually in young to middle-aged adults with the appearance of small, uncontrolled movements. The symptoms gradually increase in severity until the patient is confined to bed. Decreased mental capability, including memory loss and personality change, also occur over time, and death ensues 15-20 years after onset. Thus, the disease affects both motor control and cognitive processes.

The disease is inherited in a dominant fashion; that is, someone who inherits one copy of the gene is affected and everyone who inherits the defective gene comes down with the disease. This means that if one parent has the disease, there is a 50 percent chance that each offspring will be affected. And if one identical twin has the disease, the other is certain to get it (although such cases have not been reported, as far as I am aware).

In the 1980s, James Gusella and his colleagues at the Massachusetts General Hospital localized the gene for Huntington's disease to human chromosome 4. It took them and others another 10 years to isolate, clone, and sequence the gene, illustrating that identifying and analyzing mutated genes is not a simple task.

The gene codes for a large protein, which was given the name Huntingtin, but its function still remains unknown. The protein is found in a variety of cells in the brain and elsewhere, but why it seems to particularly affect some neurons and not others, and some neural processes and not others, is very much a mystery.

The nature of the defective gene and protein has now been unraveled. The mutant protein has an excessive number of a particular kind of amino acid near one end of the molecule. The amino acid is called glutamine and it is coded in the gene's DNA by the nucleotide sequence cytosine-adenine-guanosine (CAG). Thus in the defective gene, the CAG sequence is repeated over and over again. In normal people, there are some—between 6 and 34—CAG repeats in the gene. However, people with Huntington's disease have from 37 to more than 250 CAG repeats. There is, however, a bit of variation here. Some people with Huntington's disease have just 36 repeats and a few older people who have 39 repeats appear quite normal. But anyone with more than 40 repeats definitely gets the disease.

Observers have long recognized that there is considerable variability in the onset and progression of the disease and some—particularly those arguing that much more is involved in behavior phenomena than single genes—suggest that the variations are due to other genes or to environmental effects. Variation in the effects of a single mutated gene is often termed gene penetrance. Recent data suggest that in Huntington's disease the variability relates mainly to the structure of the defective gene.

This became clear when it was noticed that the onset, progression, and severity of the disease increased in some families over generations, especially when the disease was passed on from the father. When the defective gene was looked at in different generations of these families, the number of CAG repeats increased with succeeding generations. It was then possible to relate the severity of the disease to the number of CAG repeats, and this finding appears to explain much of the variability in the disease. The more CAG repeats in the gene—and hence the more glutamines in the protein—the earlier the onset and the faster the progression of the disease. Why the number of CAG repeats increases in succeeding generations of these families is not understood.

Several neurodegenerative diseases have now been linked to increased numbers of CAG repeats in specific genes but we don't know why this causes problems. It might be that proteins with excess numbers of glutamine molecules accumulate in the cells—the cells have difficulty in breaking down the excess and abnormal protein—and become toxic to particular neurons. Indeed, there is evidence that consecutive and excessive glutamines in a protein make it "sticky," so that one possibility is that the mutant proteins stick to each other and or to other proteins, causing an accumulation of protein. Another possibility is that fragments that result from breakdown of the mutant protein are toxic. Thus, there might be a gradual accumulation of toxic protein aggregates in certain cells, or toxic products might accumulate during protein breakdown, and both situations could lead eventually to the death of the neuron.

Complex and Cognitive Behaviors

I know of no non-disease-related behavioral or cognitive trait in humans that can yet be related to a single gene. It is generally believed that complex behavioral traits result from the products of many genes. At first glance, a promising exception is FOXP-2, the gene that causes the speech and language disorder in the large human family described in Chapter 3. The gene defect is found in every affected family member, but not in any normal family member nor, so far, in any unrelated normal person. One unrelated person who has the disorder has also been found to have the same gene defect. As noted in Chapter 3, the gene is dominantly inherited and leads to defects in speech and language, including grammatical skills.

The function of the gene is not known, although it has been proposed that it is important in the development of circuitry related to speech and language skills. Broca's area, for example, could be one possibility for the site of action of the gene, although there is no evidence for this. The important point is that, whereas this is a single gene, it codes for a transcription factor, and transcription factors can turn on and off many genes. Thus, it is likely to be not just a single gene and protein involved here, but ultimately many genes and many proteins probably playing a role.

Genes, Environment, Development, and Behavior

So far, the examples I have chosen stress the importance of genes in behaviors, but this is only one side of the story. What can we say about the other side, environment and other developmental factors on behavior?

My current research is mainly with zebrafish, small, freshwater, bony fish that are convenient for genetic studies. Zebrafish are vertebrates—animals with backbones—as are we, and they are at present the only vertebrate with which we can do certain kinds of genetic studies. We can inbreed zebrafish so that they become virtual clones. In other words, the genetics of the fish we use are essentially identical.

My laboratory is interested primarily in the visual system, especially the retina. We produce mutations in the fish's genome by altering the genetic material in the fish with specific chemicals and then we look for changes in the structure, function, and development of the retina. We find single-gene mutations that affect the retina fairly frequently; many of these alter the visual behavior of the animals, not unexpectedly. Disrupting some aspect of the function of the light-sensitive cells in the retina, the photoreceptors, or other retinal neurons, obviously decreases the animal's ability to see properly, and this affects visual behaviors. For example, we have found single gene mutations that render animals completely blind, partially blind, color-blind, or movement insensitive. In many cases we have identified the genetic nature of the defect, but again this is not at all surprising.

However, we have also made a number of observations that bear on the question of genes, environment, and other possible factors on higher-order behaviors. Let me describe one inadvertent experiment carried out by a young colleague. He was interested in observing zebrafish in a tank and began with 12 young adult fish, all of which came from the same cross. That is, they were all siblings, from the same parents, of the same age and raised under identical conditions. Genetically they were as similar as zebrafish can be and they were raised under as identical conditions as is possible—in the same tank, at the same temperature, and so on.

To make his observation tank more interesting, my colleague put an elongated plant at one end. Shortly thereafter, as he was observing the fish, he noticed that one fish was swimming in the open, whereas the other 11 were staying in or close to the plant. Curious, he disturbed the surface of the water in the tank, which usually indicates to the fish that food is coming. The fish charged out from the plant and milled around looking for food. There was none, but then my colleague noted a most interesting phenomenon. The fish that was originally swimming in the open began to chase the others back into the plant. This took a few minutes, but he or she eventually succeeded. It was clear that this fish was running the show!

We speak of this behavior as dominance and it is commonly observed in fish and many other animals. The interest here is the fact that all the fish were essentially identical genetically—not

100 percent identical, but close to that. Genetic factors presumably can be ruled out as playing a role here. We might suppose that it is an environmental effect, but here, too, we are stumped. The animals were all raised identically. There were some size differences among them, but that didn't seem to be key in follow-up experiments as to who would be the dominant fish. The answer is still not clear, but this example makes the point that either extremely small genetic differences or unrecognizable environmental differences can make a big difference in an animal's behavior.

But possibilities other than genetics or environment might also be at play here. Chance, for example, could play a significant role. That is, during brain development many things go on—cell specification and differentiation, cell migration, axonal pathfinding, pruning of neurons, rearrangement of synapses, and so forth (Chapters 1 and 2). Some chance variation in these processes in genetically identical animals raised under identical conditions is likely to occur and this could give rise to significant behavioral differences. The bottom line is that it can be very difficult to cull out behavioral, genetic, or other (developmental) differences among organisms and to assign these differences to specific behavioral traits.

Another set of factors that might be involved here are the so-called "epigenetic" factors that regulate how genes are expressed in cells and how much protein is made. We know, for example, that gene expression can be modulated by enzymes that add methyl groups onto the DNA bases, or how the DNA molecule is made accessible for the expression of its genes. In the nucleus, the long DNA molecules are wrapped tightly around specific proteins to form a highly condensed structure called chromatin. There are enzymes, chromatin-remodeling complexes, that unwrap portions of DNA from chromatin to make it accessible to transcription and other factors that lead to expression of the genes on that bit of DNA. We still know little about these epigenetic factors, but obviously variations in them can alter gene expression quite significantly. In other words, it is not simply gene sequence—that is, the particular protein produced from a specific gene sequence—that is important, but how much protein product is produced from a specific gene, and this amount can be influenced by epigenetic factors.

Twin Studies

What has long been considered the gold standard for uncovering the relative influences of genes and environment on behavior in humans is the study of identical twins reared apart. However, this field has been fraught with controversy, including claims of scientific fraud against one of its pioneers, the Englishman Cyril Burt, who studied identical twins reared apart from the 1920s to the 1960s. And at least some of the claims of fraud against Burt appear to be true. However, over the past 20 years, the field has been revisited by Thomas Bouchard and his colleagues at the University of Minnesota, and their work is careful, detailed, and compelling. It is based on a study of more than 100 sets of reared-apart twins or triplets, mainly from the United States and the United Kingdom, but twins from six other countries were examined as well.

The twins were brought to Minneapolis where they were put through 50 hours of medical and psychological testing over the course of a week. In some cases the twins had had contact with one another over the years, but in others they did not. In some cases, the twins were raised in families with quite similar backgrounds and values, but in others they were not. Regardless, the main finding of the study was that identical twins raised apart are about as similar as identical twins reared together. The conclusion drawn was that being reared in a different environment does not appear to influence many behavioral traits and therefore genetic factors must play a significant role in determining a variety of behaviors.

This is not to say that environmental or other factors do not play a key role in determining human behaviors. For no trait examined was the result identical among the pairs of identical twins, whether they were raised together or apart, and the correlations ranged from low (0.34) to high (0.78). (These numbers mean that as little as 11 percent to as much as 61 percent of the variability is genetic—the correlation is squared to arrive at the variance.) Although it is difficult to put hard numbers on all of this, the conclusion drawn is that about 50 percent of the variation in personality traits is genetic; the other 50 percent is due to environmental or other factors. It should be noted, however, that many psychologists believe an heritability estimate of 50 percent for personality traits in humans is too high, based primarily on animal (rat) studies in which heritability of behavioral traits has been shown to be much lower (<0.2).

The big surprise at first glance from the twins study was that the home environment did not have more effect on behavioral traits. There are two comments to be made here. First, the rearing environments were probably not all that different among the cases of reared-apart twins. We raise children in much the same way throughout the developed world, and all the twins came from the developed world. It is when a rearing environment is deficient and children are deprived that we see substantial differences in behaviors that can be permanent.

The Romanian orphans are a case in point. As a result of harsh laws banning birth control and abortion in Romania in the 1970s, more than 100,000 unwanted children were born but then abandoned in orphanages. They were given little environmental stimulation and had little personal interaction with caring adults. After the fall of the Ceausescu regime, many of these children were rescued and placed in adoptive homes in the West. Those adopted at younger ages (before six months) have done reasonably well, but children adopted at older ages (one to three years) have shown persistent personality defects into their teenage years. The guess is they will never recover emotionally or socially as a result of that early deprivation.

The Romanian orphan situation is reminiscent of the experiments of Harry Harlow of the University of Wisconsin in the 1950s and 1960s. Working with infant rhesus monkeys, he showed that these monkeys often demonstrate lasting adverse effects if deprived of social and maternal interactions early in life. (Other monkey species, on the other hand, are less affected by social or maternal isolation.) Rhesus monkeys show depression and other behavioral disturbances when isolated from their mothers and other monkeys at an early age, and often these are permanent effects, although if deprived monkeys are subsequently placed with normal infant monkeys, they do appear to recover a great deal of normal behavior. In some experiments, the infant monkeys were housed in close proximity to other monkeys, both young and old, so that they could see, hear, and even smell them. Nevertheless, the monkeys deprived of contact with other animals showed severe behavioral disturbances. Physical interaction appears to be important in the nurturing of many young primates and other mammals. Harlow showed, for example, that maternally deprived infant monkeys would cling to an artificial terry-cloth mother and ignore a wire-frame mother, even if the wire-frame mother was the source of food—its milk bottle was attached to the wire-frame mother.

Of interest in this regard is the recent finding that mouse pups that lack a receptor protein activated by opiates show much less distress when isolated from their mothers as compared to normal pups. This suggests that during bonding between mother and infant, endogenous opiate-like substances are released in the infant's brain, activating the receptors and leading to pleasurable feelings that contribute to the bonding between mother and child. Opiates are well known to alleviate pain as well as provide pleasurable feelings, and thus the lack of activation of opiate receptors in the brains of socially isolated animals could, perhaps, result in painful feelings that contribute to the distress shown by such infants.

Other experiments with rats have shown that newborn animals that receive no touch stimuli during an early critical period show increased levels of the stress hormone cortisol in response to various challenging situations that the animal encounters even when an adult. Excessive levels of cortisol have also been found in the Romanian orphans when they were subjected to a somewhat stressful event—being examined by a doctor unknown to them, for example. As was noted earlier in the chapter, increased levels of the stress hormone leads to a decreased proliferation of neurons in the hippocampus, and there is also evidence that increased levels of the stress hormone cortisol cause brain damage in animals, including neuronal cell death. In aging humans, increased cortisol levels have been linked to difficulties in performing certain memory tasks. (Chapter 6 discusses this further.)

The second comment with regard to the apparent lack of effect of home environment is that children themselves play a significant role in the type of environment they experience growing up. They select the parent to whom they respond more closely, their friends, and the activities they most enjoy and prefer to do. Children demonstrate different temperaments, and parents usually respond predictably to these temperaments. One doesn't hug excessively a child who obviously dislikes hugging. I am not saying, and others have made this point in much more detail and more persuasively than I do here, that the home environment is not important; it certainly is, as the experience of the Romanian orphans so clearly demonstrates. However, as long as the home environment is supportive and caring, genetic factors play an important role in the determination of behavioral traits.

IQ and Genetics

Of all the topics touched upon in this book, perhaps none is more contentious than IQ and inheritance. The topic generates much more heat than light. Nevertheless, it is one that cannot be entirely ducked in a book that deals with nature-nurture issues.

The first question that might be raised is whether the concept of IQ has validity. Can intelligence be realistically and reliably measured and given a number, usually termed "g"? Everyone agrees that many cognitive skills are not included in present IQ tests, so what does an IQ score really mean? There is no question that IQ does predict such things as success in school, job performance, and even future economic and social status, but, and this is a very big "but," not perfectly. And it is the big "but" in the last sentence that is the hang-up—how well does IQ predict these things, and is too much left out in determining a single "g" number for someone?

A number of psychologists, especially Howard Gardner of Harvard University, espouse the idea of multiple intelligences, and I find this notion attractive. When I compare my wife's intellectual abilities to my own, I am struck by her obvious superior abilities in some areas compared to mine, and I believe I might have some abilities superior to hers (although I dare not tell her so). Gardner has proposed seven categories of intelligence: linguistic, musical, logical-mathematical, spatial, bodily-kinesthetic, intrapersonal, and interpersonal. One of the strengths of this categorization—that gives it some neurobiological meaning—is that each form can be selectively affected by one or another type of brain damage, as Gardner points out. Others have proposed somewhat different categories, but the idea is basically the same—that we all differ behaviorally and cognitively in different ways and this should be taken into account in evaluating intelligence.

However, at the moment we are stuck with one IQ score and the question is how much of IQ can be attributed to heredity, to environment, or to other factors. Needless to say, these are complex relationships that are not easily teased apart. Bouchard's twin studies indicate that the variance in IQ due to genetic variation is high—60-70 percent. Earlier studies had suggested somewhat lower values—40-50 percent—with regard to the heritability of IQ, but Bouchard's results were obtained from middle-aged adults, and recent findings suggest that the correlation of IQ and inheritance increases with age. Indeed, there is good evidence that IQ is more environmentally linked in children, youngsters, and young adults than in middle-aged or older people. For example, IQ scores in children correlate with the education of their parents in every society studied.

One can also correlate lower IQ levels with lower birth weights. Alcohol and drug use as well as cigarette smoking during pregnancy also seem to decrease IQ levels. Lead exposure both in pregnancy and in young children also leads to lower IQs, as does a deprived environment during the early years. It is the deprived-environment children who seem to benefit most from social programs such as Head Start and whose IQs appear to be raised significantly by the programs. But here again, the data are somewhat murky and often disputed. The effects of programs such as Head Start seem to peter out as children get into school. Schooling itself appears to influence IQ; each year of schooling might increase IQ by two to four points.

The bottom line here is that it is the most deprived children whose IQs can be raised most by environmental intervention. The IQs of children raised in solidly supportive and caring environments are less environmentally related. But exactly how much the "less" is is hotly debated, and neurobiology can contribute nothing to the debate.

I end with a most curious observation with regard to IQ scores that needs, in my view, much more examination. And this is the fact that IQ scores are gradually rising—at a rate of about three points per decade as shown in Figure 5-3.

What might this mean? Are we becoming more intelligent or just better at taking the tests—training our children to deal better with them. The rise of IQ scores over time was summarized in the mid-1980s by James Flynn in New Zealand and is often known as the Flynn effect. The increase is substantial: In 1932, about 2.5 percent of the population scored above 130 on the IQ test and were categorized as in the "very superior" group. By 1997, about 25 percent of the population would score at this level if they were to take the 1932 test. Has intelligence risen by that much? How could that be possible?

Both sides of the issue are argued. Some suggest that we are smarter because we are bigger and that means a bigger brain. In the industrialized world, we have been gaining an average of 1 cm in height per decade, thought to be due to better health and nutrition. The implication here is that we are smarter than our parents and our children are smarter than we are. Some are comfortable with this notion, but others are not.

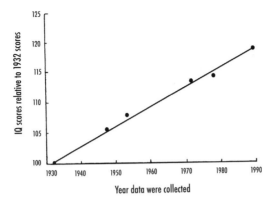

FIGURE 5-3 The Flynn effect: There has been a consistent increase in mean performance on a variety of IQ tests from the early 1930s to the present. Why this is so is not understood.

On the other side are those who insist that it is an environmental effect—schooling and training children to take the tests is one possible answer, but has schooling changed enough since the 1930s to make this much of a difference? Schooling relates more to content than reasoning and it is in reasoning that the test results show the greatest increases.

An interesting factor that has been proposed and is clearly worth exploring further is the visual environment to which our children are now exposed compared to those we or our parents experienced. Clearly, the visual environment available to us is more complex and has become increasingly more so since the 1920s when movies were introduced. Television was another milestone enhancing the richness of our visual world, and then computers and computer games arrived in the 1980s. Of course, the computer introduces more than an enriched visual environment, so it will be interesting to see if IQ scores tend to rise slowly over the next few decades or, indeed, rise even faster. There might be neurobiological implications here if it becomes possible to determine the effects of such enriched environments on brain structure and/or function.

An intriguing study recently published in the journal *Nature* bears on this issue. When subjects between the ages of 18 and 23 who played action video games for at least one hour per day four days per week for the previous six months were tested on visual attention tasks, they performed significantly better than subjects the same age who did not play such games. For example, the video-game players were superior in determining how many objects were present in a briefly flashed display. The differences in performance by the video-game-playing group compared to the nonplaying group were highly significant. Improvement in such tasks by the nonplaying group was achieved by having them play a specific game for one hour per day for 10 consecutive days. These findings convincingly demonstrate the effects of training on certain cognitive tasks.

PART III

THE AGING BRAIN

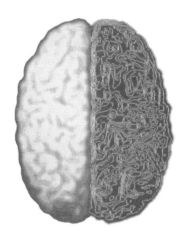

IS AGING OF THE BRAIN
A DISEASE?

We have all witnessed the deterioration of our parents or grand-parents as they aged. Too often, it is a sad deterioration with a loss of quality of life as they fail mentally. At the same time, we read about the marvelous advances in medicine that are supposed to make us healthier and live longer. What is going on? Are people living longer? And if so, what are the consequences? Is it likely that we could live to be 150 or even 200 years old, and what would it take for us to achieve that goal if, indeed, this is something we would want to do?

These are the topics I cover in this chapter. I will first describe what seems to be happening in the brain as we age. But this topic, like many others discussed in this book, generates much contro-versy and the neurobiological facts are disputed. Nevertheless, there are a number of things that can be said.

When I was in medical school, no one talked about Alzheimer's disease in people over 65 years of age; at that time the failing mental abilities seen in those over 65 were described as senile dementia and a natural consequence of aging for many. Today the view is different. Alzheimer's disease is thought of as a real disease regardless of the age when it strikes. But is this an accurate perception?

A number of other age-related degenerative diseases such as Parkinson's disease and age-related macular degeneration can occur as people age. What causes these diseases? How might we combat them and what types of therapies are possible? Here I shall use as a model an inherited retinal degeneration, called retinitis pigmentosa, which typically attacks people in their twenties and progresses until complete blindness ensues by age 60 or so. It might be the best-understood neurodegenerative disease at the moment, and therapies for its cure, at least in animals, look promising.

Neuronal Changes with Age

I noted in Chapter 1 that virtually all neurons are generated in the developing brain by about six months of age, but the brain is then only about 30 percent of its adult size. The brain grows rapidly in the first three years of life, but does not reach its maximum weight until about age 20, as Figure 1-3B showed. Much of that increase is due to the growth and elaboration of neurons, shown in Figure 1-4, and presumably neuronal circuitry. After age 20, on the other hand, brain weight and volume gradually decline and this continues for as long as we live.

How much brain weight is lost and what does the loss mean? Everyone agrees that the brain shrinks with age—this is seen in both humans and other primates. The figure often given for humans is up to 15 percent shrinkage over a life span of 100 years, but some investigators believe this figure is too high. But what does a loss of even half of 15 percent mean?

Because few new neurons seem to be generated in the adult brain (as discussed in Chapter 5), one possibility is that the shrinkage reflects loss of neurons, and certainly this occurs. If the shrinkage were due entirely to neuronal loss, the number of neurons lost per day would be astonishingly high—about 200,000 per day or more than 8,000 per hour! This calculation is based on an estimated volume loss of 7 percent over a life span of 100 years with the number of neurons initially present (at one year of age) being 10^{11} or 100 billion (a conservative estimate).

In the 1950s, Harold Brody of the State University of New York examined the brains of 20 human subjects from newborn to age 95 and reported extensive cell loss—up to 40 percent in several regions of the cortex. However, later studies questioned these results, and some researchers reported either minimal neuronal cell loss or even none in many areas of the cortex. One reason suggested to explain Brody's results was that cases of Alzheimer's disease or other dementias might have been included in his sample, and it is well known that these conditions cause massive cell loss.

There is a general agreement that brain shrinkage occurs, so if neuronal cell death is not the major contributor, at least in the cortex, what is going on? Neuronal atrophy is one possibility and there is substantial evidence that this happens with age. Just as neurons in the very young brain grow larger and extend more dendritic branches that go longer distances (shown in Figure 1-4), neurons in the aging brain do the opposite—they shrink and have fewer branches. The atrophy of neurons and neuronal branches almost certainly results also in fewer synapses in the aging brain.

Another factor is loss of white matter in the brain—those regions where axons are most prevalent—and again, there is both human and animal evidence for this. Indeed, recent studies have reported losses of as much as 30 percent of the white matter in aged brains. One suggestion, for which there is increasing evidence, is that the myelin sheaths surrounding axons break down with age. This could account for the decrease in white matter volume and also for some of the cognitive changes that occur with age. For example, myelin promotes more efficient transmission of the electrical signals that travel down axons, and if this transmission is slowed, the ability to process information could also be slowed. And, as I shall describe below, older people process information more slowly.

Biochemical measurements also show significant changes

with age. Total brain protein is down by as much as 30 percent in 80-year-old brains compared to protein in the brains of 20-year-olds, and NMDA receptors have been reported to be lower by about 30 percent in aged brains. This finding might partially explain one of the most common complaints of older people—a loss of memory recall. (Remember the importance of NMDA receptors in memory and learning that was discussed in Chapter 4.)

What can we conclude from present data? Whereas it is probably correct that there are regions of the brain, including certain cortical areas, that show minimal neuronal loss with age, certain other regions—for example, some subcortical areas, the brainstem, and other regions—do show substantial cell loss, as much as 40-80 percent. In the retina, for example, careful measurements of the photoreceptors, which are quite easy to count accurately, indicate a loss of 30 percent of both rod and cone photoreceptors from teenage years to age 80.

So, many brain regions certainly have substantial cell loss, and all regions are likely to have some. Neuronal loss might also be selective for certain types of neurons. Betz cells, which are very large neurons found in the motor cortex, seem to wither away with age and are pretty much gone by age 80. And white matter, NMDA receptors, and proteins in general seem to be lost. A caution, noted above, is that the dementias are so common in elderly people, especially over age 85, that a few diseased brains might have been included inadvertently in some of the studies these above conclusions are based upon. Conversely, brains that were found to have substantial neuronal cell loss might have been excluded from some of these studies because of a belief that they are diseased—brains from people with early Alzheimer's, for example.

Cognitive Changes with Aging

Everyone over 50 complains of not remembering things as well as when younger, and it is certainly true, as shown by cognitive studies, that older people have decrements in learning and memory. Not all types of memory seem to be equally affected. Delayed memory—recall of something learned a while ago, for example, remembering a name—is much more affected than immediate memory—recall of something that has just happened—but with age more time is needed to learn new information and it is harder to stay focused on a memory task. When different brain regions were examined to see if shrinkage of a specific area could be correlated with the loss of memory function in people between ages 55 and 85, the only area that showed a significant decrease in volume (measured by brain imaging techniques) was the hippocampus. We shall return later to the issue of memory loss and hippocampal changes.

Other cognitive deficits have been seen in older people. For example, the manipulation of information slows down, so it takes an older person longer to come up with a response to a complicated scenario. This deficit is thought to be related to changes occurring mainly in the frontal lobes, which are concerned with reasoning, planning, and keeping things in one's mind. Older people need to reflect more on an issue than do younger adults. On the other hand, vocabulary and language skills do not appear to deteriorate much, if at all, in normal people as they age; neither is IQ or abstract thinking much affected.

When do these changes begin to take place? Already in one's twenties, changes appear to be occurring. For example, psychologists can measure differences between 30-year-old subjects and people in their late teens or early twenties. Information is processed a bit slower and is held for a shorter time in conscious awareness, and recall is somewhat less efficient in the 30-year-olds. Cognitive scientists agree that aging of the brain begins at about the time brain volume begins to decline, and this decline of cognitive abilities and brain volume continues for the rest of our lives.

How Long Could We Live?

It is common knowledge that average life expectancy has increased spectacularly in the past 100 years. In Europe and the

United States, the average life span was less than 47 years in 1890 and by the 1990s it was more than 75 years. During the decade of 1968-1978, average life expectancy rose at the phenomenal rate of one month per year for all those over 50! In Japan, the figures are even more impressive; by the mid-1990s the average life expectancy for women was about 83 years of age. Japanese men, like men the world over, had a lower average life span of 77 years. Developing countries also showed substantial increases of average life expectancy in the 1990s.

But what about absolute life expectancy? Has that increased? Here the news is quite different. Ancient texts mention individuals living to 120 years of age, and today we occasionally hear of someone that old, but this is very exceptional. The age at death of the longest-living human that is well documented was 122 years, and in the fall of 2003 the then-oldest man in the world died at the age of 114. He was Japanese and the oldest woman alive at that time, also Japanese, was 116 years old. Indeed, if one looks at the trends in human longevity from antiquity to the present day, it has not increased significantly if at all, as shown in Figure 6-1.

Whereas average life expectancy has increased dramatically

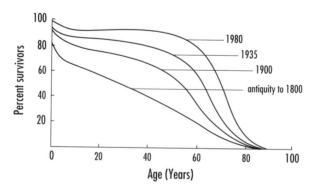

FIGURE 6-1 *The changes in life expectancy from ancient times to 1980. Average life expectancy has changed dramatically from about 35 years 200 years ago to more than 75 today. Absolute life span of humans has not increased significantly since antiquity. Only a small percentage of humans live to be more than 100.*

as a result of medical advances and improved housing and sanitation, the maximum life span has not. The number of people who live to be 60 has increased from less than 20 percent in the early nineteenth century to more than 80 percent today, but the endpoint of life expectancy remains at about 100 years of age.

Although the average life expectancy has been increasing virtually linearly since the 1960s in Western countries, it is expected to level off with an average life expectancy of about 85. In other words, it is likely that we are already coming close to our maximum average life expectancy, if there is a biological limit to absolute life expectancy.

My view is that this limit is real and I suspect it might relate to the brain and its aging. Whereas we can replace hearts, lungs, and livers, we cannot replace brains or even brain cells, at least at the moment, and some believe that we will never be able to replace whole brains. Indeed, as someone glibly pointed out, if whole brain transplants were possible, it would be far better to be the donor rather than the recipient, for obvious reasons! This is why there is so much interest in the possibility of stem cells remaining in the adult brain (discussed in Chapter 5). Indeed, if they are present generally, or even in a relatively few places, and could be induced to generate a variety of new neurons to replace dying or dead ones, one might suppose that we could renew our brains and increase maximal life span.

The transplantation of embryonic stem cells into a brain to replace dead neurons and maintain brain circuitry is another possibility that is receiving much attention. Alternatively, it might be possible to find ways to stop or slow the neurons' aging processes. All these possibilities are being explored, but at the moment they are still very distant.

But do we even want to extend our life span significantly? Perhaps this is a philosophical question; nevertheless, it has biological implications. Evolution depends on organisms having finite lifetimes, so that different mixes of genes can be expressed in new generations. But are humans still evolving biologically? I am not sure anyone knows.

If we did live 50 percent longer, to 150-180 years, population pressures would be substantially increased, and overpopulation on Earth is already an enormous problem. Would we be willing to restrict population growth to maintain a habitable planet? And, of course, the presumption in all of this is that quality of life would be maintained at a vigorous level for much longer, so retirement would be at 130-150 years of age, rather than between 60 and 70 as it is now for most people in the developed world. How would this affect human creativity, which depends so much on the young. What might be the economic implications if people lived for 150 years? None of this has been thought through. Perhaps it should be.

Alzheimer's Disease

No disease of the elderly is more feared than Alzheimer's disease. It strips sufferers of dignity, and eventually they lose those traits that make them unique human beings. As many as 4 million Americans suffer from the disease, and one estimate suggests that by 2040 as many as 14 million people will be affected in this country alone.

Those who suffer from it commonly show a decline in mental abilities in their late fifties or early sixties, beginning with deficits in recent memory, and progressing to a loss of virtually all higher mental functions. Confusion and forgetfulness are the most common symptoms, followed in some patients by difficulty in executing motor acts that previously were simple to do, and even loss of speech. A few people have a much earlier onset of Alzheimer's disease—but many of these come from families in which the disease is inherited as an autosomal dominant disorder of midlife. Many more people show Alzheimer's-like symptoms as they age and eventually display the full-blown mental deterioration of the condition. One study showed that at age 65 about 15 percent of the U.S. population is showing Alzheimer's-like symptoms and by age 85 as much as 50 percent is showing them.

But, you say, my Aunt Marian lived to be 102 and she was as

sharp as a tack almost to the day she died. This is certainly correct, that certain people live to be 100 or a bit more and show virtually no symptoms of the disease. But, unfortunately, it is a relatively rare person who reaches the century mark without evidence of one or another neurodegenerative disease. Alzheimer's accounts for about 50 percent of the neurodegenerative diseases, and changes found in the brains of Alzheimer's victims are often found also in the brains of "normal" aged people, although in them they are much less severe.

The striking feature of the brains of Alzheimer's patients is a tremendous loss of brain volume and massive degeneration of neurons. Significant portions of the cortex atrophy, and PET scanning indicates a marked reduction in brain metabolism. One region of the brain particularly hard hit is a nucleus found on the basal surface of the forebrain, called the nucleus basalis. Most of the synapses that release acetylcholine in the cortex are made by the axon terminals of neurons coming from nucleus basalis neurons. In Alzheimer's patients as much as 60-90 percent of the enzyme responsible for synthesizing acetylcholine in the brain is lost. One treatment being tried for the disease is an inhibitor of the enzyme that breaks down acetylcholine in the brain, the idea being to raise acetylcholine levels. Its success so far is modest at best. The neurons in the nucleus basalis are thought to play a role in integrating both subcortical and cortical information processing, and thus, degeneration of this nucleus could be key in explaining the cognitive defects in Alzheimer's disease.

Why do neurons deteriorate and eventually die in Alzheimer's patients? Neuroscientists are beginning to get a handle on the answer. As was the case for Huntington's disease, it appears that the accumulation of toxic proteins causes the neurons to die. But unlike Huntington's disease, the toxic proteins that accumulate are extracellular (outside the neurons) in Alzheimer's disease. The principal protein that accumulates is β-amyloid, a relatively small protein consisting of 40-42 amino acids that is a natural constituent of the extracellular space in the brain and tends to build up in all of us as we age. However, a prominent feature of brains affected by Alzheimer's is the presence of many dense accumulations of protein, mainly β-amyloid, in prominent extracellular structures called plaques as shown in Figure 6-2.

The aggregates form because β-amyloid is sticky, especially when it is somewhat misfolded. This occasionally happens with proteins, which are, of course, long chains of amino acids that must be folded correctly to carry out their specific function. In forming plaques, β-amyloid molecules stick to each other and to other proteins, forming a mass that binds to both neurons and glial cells. Having a glob of sticky stuff on a neuron is probably enough to disrupt its function and structure, but the immune system also comes into play. The plaques set off an inflammatory reaction that further complicates matters in the region and results in more cell death and brain damage.

Plaques and β-amyloid are not the only things that accumu-

FIGURE 6-2 The histological changes seen in the brains of Alzheimer's patients. The arrow in the lower left points to a plaque that consists mainly of misfolded β-amyloid protein encompassing neurons and presumably disrupting their function. The other arrows point to neurons filled excessively with another protein called tau.

late in the brains of Alzheimer's sufferers. An intracellular protein called tau also accumulates, forming tangles inside neurons. Present evidence suggests that the accumulations of tau in tangles, as well as the loss of neurons, is secondary to the plaque formation. Indeed, it might be that the tangles are a sign that a neuron is dying. On the other hand, in a variant of Alzheimer's disease called Pick's disease—which has virtually the same symptoms—intracellular tau protein accumulations are the main pathological feature. Thus, accumulation of tau protein might play a role in causing neurodegeneration.

How does genetics fit into all this? At least five genes predispose people for Alzheimer's disease. At the moment, we can account for about 50 percent of the Alzheimer's cases based on a genetic predisposition (excluding those cases of early onset due to an autosomal dominant mutation). What is meant by a genetic predisposition? People with a predisposing gene might not get the disease, but the chances are higher than if they do not have that specific gene variant. For example, relatives of people with Alzheimer's who have the variant gene have an increased risk of about 38 percent for the disease at age 90. However, about 50 percent of the population shows Alzheimer's-like symptoms at that age, so of the healthy 50 percent, if they have Alzheimer's in the family, about 40 percent, or two out of five, will eventually come down with the disease. The other three will stay healthy even with the predisposing gene.

What can be said about the nature of the predisposing genes for Alzheimer's? Without getting into the details of their specific names or explaining their exact function, all appear to crank up levels of amyloid in the brain. One of the genes codes for the amyloid precursor protein (APP) from which β-amyloid is formed. This predisposing gene might have a small variation (called a polymorphism), which means that the β-amyloid protein it codes for is slightly altered. For example, more often than normal the protein might not fold quite right or perhaps it is more resistant to degradation. Both of these hypothetical scenarios would lead to more β-amyloid accumulation.

Other predisposing genes are involved in the breakdown of APP to β-amyloid and they, too, could result in more β-amyloid accumulation. Yet other predisposing genes seem involved in the breakdown of β-amyloid itself. The changes in the genes are presumably not all that large, and thus their effects are minimal so that it takes decades for the buildup of the amyloid to reach levels high enough to cause Alzheimer's.

An important point to reiterate is that β-amyloid accumulates in all of us as we age, and many brains of normal older people even show plaques at autopsy. So the key is how much and how fast amyloid accumulates. And, of course, one could have variations in these or other genes that could predispose one to *not* get the disease. Having a gene that codes for a slightly more efficient enzyme that breaks down β-amyloid is certainly possible in certain people—the Aunt Marians, perhaps, who live to 102 with virtually no cognitive deficits.

Of interest, and perfectly understandable, is the fact that three of the same genes are involved in families that inherit Alzheimer's as an autosomal dominant disorder that starts at a much earlier age. These families must have a more severe alteration in the genes, which then leads to a much more rapid accumulation of β-amyloid and the midlife onset of the disease.

Parkinson's Disease and Other Age-Related Neurodegenerative Diseases

Parkinson's disease is another age-related neurodegenerative disease commonly seen in older people. Indeed, Parkinson's disease is the second most common neurodegenerative disease, affecting about 1 million Americans over the age of 55. It is also due to degeneration of neurons, but in this case to neurons in a specific midbrain nucleus that innervates a large part of the basal ganglia complex. The basal ganglia are concerned with the initiation and execution of movement. Thus, patients with Parkinson's disease show specific motor deficits. Initially the disease is marked by a rhythmic tremor of the limbs at rest, but then it progresses to a rigidity of limb muscles. Eventually, Parkinson's patients have difficulty initiating movements. They tend to shuffle as they walk and have a mask-like facial expression. Parkinson's disease sufferers can also develop a dementia.

The neurons that die in Parkinson's disease release the neuromodulator dopamine in the brain, and a deficiency of dopamine seems to be a major cause of the symptoms. Thus, one therapy consists of giving patients a molecule that is a precursor to dopamine—a molecule that the brain can easily convert to dopamine.

This molecule, L-dopa, is made naturally in the brain from the amino acid tyrosine and with one enzymatic step can be converted to dopamine as shown in Figure 6-3.

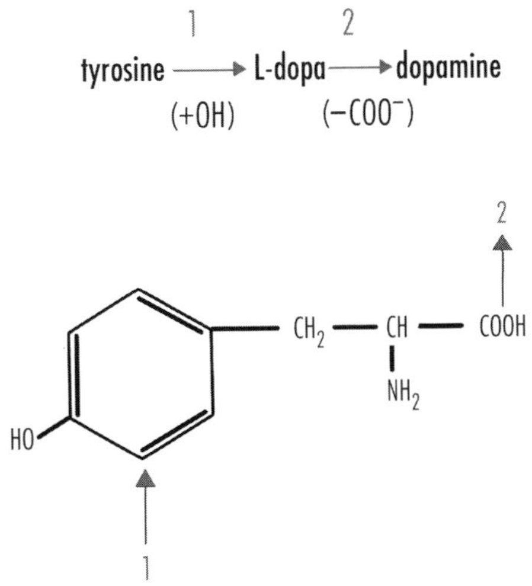

FIGURE 6-3 The conversion of the amino acid tyrosine into dopamine requires just two steps and involves an intermediate molecule, L-dopa. The first enzymatic step involves adding a hydroxyl molecule (+OH) to tyrosine at the position shown by arrow 1. The second step is the elimination of a carboxyl molecule (–COO⁻) from position 2. L-dopa readily crosses into the brain when given to a patient, but dopamine does not; hence L-dopa is given to patients suffering from Parkinson's disease.

L-dopa therapy doesn't cure the disease, but it substantially relieves symptoms in many patients for substantial periods. It does not work for all because it causes side effects in some patients. Another treatment, much riskier and more experimental, is the transplantation of dopamine-secreting cells into the brain. This was described briefly in Chapter 5, where I pointed out that this approach has so far met with only modest success.

How much neurodegeneration must occur for the symptoms of Parkinson's disease to occur? This has been quite convincingly documented, and the findings indicate there must be a loss of about 50 percent of the dopaminergic neurons themselves and a decrease in dopamine content in the basal ganglia of about 75-80 percent before the classic symptoms appear. Thus, a substantial amount of neurodegeneration must happen before the disease becomes apparent.

The causes of Parkinson's disease are not well understood, although some suggest that it, like Alzheimer's and Huntington's diseases, might be caused by the accumulation of a toxic protein. Cytoplasmic inclusion bodies, called Lewy bodies, are found in the dopaminergic neurons early in Parkinson's disease. One protein, called α-synuclein, is the major component of the Lewy bodies. The function of α-synuclein is not well understood, though it seems to have a role in synaptic function and synaptic plasticity and might be involved in the release of dopamine at dopaminergic synapses. The α-synuclein molecule, like β-amyloid, is sticky and binds to other α-synuclein molecules as well as to other proteins.

With Parkinson's, as with Alzheimer's, several genes, including the gene that codes for α-synuclein itself, have been implicated as playing a predisposing role. Two mutations in the α-synuclein gene have been linked to rare forms of Parkinson's disease, and these mutant genes as well as the normal α-synuclein gene have been expressed in the fruit fly, *Drosophila melanogaster*, using transgenic techniques. Interestingly, all three expressed genes lead to a Parkinson's-like disease in fruit flies, causing Lewy bodies to appear in the dopaminergic neurons and producing a late-onset locomotor dysfunction in the flies. There is, however, little difference in the effects of the three expressed proteins, indicating that normal α-synuclein is almost as toxic as the mutant protein. This agrees with studies on humans in which it has been shown that most Lewy bodies in Parkinson's patients are made up of normal α-synuclein molecules.

A number of key questions remain. Why does α-synuclein accumulate, why does it accumulate specifically in the dopaminergic neurons, and how does it kill the dopaminergic neurons? And also, what are the triggering events for Parkinson's disease? There are clearly predisposing genes for some instances of Parkinson's, but most cases seem not to be explained in this way.

Environment and Neurodegenerative Diseases

The emphasis in the discussion so far has been on the role of predisposing genes in the causes of age-related neurodegenerative diseases. But many cases of these diseases cannot be linked to a genetic factor, so it is believed that environmental factors must be the precipitating cause. Only about 50 percent of the cases of Alzheimer's disease can be linked to one of the five predisposing genes for it. The cause of the other 50 percent is very much up in the air. The genetic links to Parkinson's disease are even weaker.

Clearly, brain injury or trauma can lead to excessive brain cell loss and people who have had such injury can show Alzheimer's-like symptoms. People prone to brain injury, like prizefighters, can develop these symptoms and in such cases are described as punch-drunk. Suffering from a prolonged high fever also can cause substantial neuronal death and result in an Alzheimer's-like condition. Such brain cell death might not be distributed evenly throughout the brain, but confined to a specific region or set of cells for some unknown reason. The former heavyweight boxing champion Muhammad Ali might be an example of this. He is said to be suffering from Parkinson's disease. He shows a severe tremor, walks sluggishly, has difficulty initiating movements, and displays an expressionless face. It seems quite possible that his disease is linked to the pounding his head took during his fighting days.

Toxins and toxic substances can also be involved in some cases. A chemical called MPTP, discovered because it was formed as a by-product of defective heroin synthesis by drug dealers in California, led to tragic consequences in a number of young drug users. MPTP breaks down in the body to a neurotoxin, MPP^+, which selectively destroys dopaminergic neurons. Those unfortunate to take such defective heroin show symptoms of severe Parkinson's disease within a few days of ingestion. Other toxins have been shown to induce Parkinson's-like conditions, so it is possible that both man-made and natural toxic substances can be involved in the disease's etiology.

Parkinson's disease is also linked with the great influenza epidemic of 1917. A subgroup of people who survived the epidemic subsequently came down with severe Parkinson's disease. Here the precipitating cause might have been a virus. These patients were some of the first to be treated with L-dopa by the famed neurologist-author Oliver Sacks. Sacks subsequently wrote a book on this experience, called *Awakenings*, that was turned into a well-received film.

Finally, the question has been raised as to whether excessive stress can predispose to neurodegenerative disease. When an animal or human is stressed, one of the things that happens is an increased release of glucocorticoids from the adrenal glands. In the short term glucocorticoids are helpful to a stressed animal. Among other things, they promote the breakdown of protein to glucose, helping to make fat available for use and increasing blood flow. With prolonged stress and a prolonged release, the glucocorticoids can have damaging effects, including increased blood pressure, gastric ulcers, and depression of the immune system.

Of particular interest here is the fact that prolonged stress in animals can also cause brain damage, especially to the hippocampus. In Chapter 5, I noted that glucocorticoids decrease the generation of new neurons in the hippocampus, and this can account for some of the effects of prolonged stress. However, hippocampal volume is also reduced in prolonged stress and one study showed a loss of hippocampal CA1 cells in stressed rats.

The situation with regard to stress and the aging human brain is by no means clear, but some studies on aging people suggest a relation between stress hormones and difficulties with memory tasks. In one study, 11 healthy subjects in their sixties or seventies were followed for four years. Of these, six showed increased levels of cortisol, one of the main glucocorticoids released from the adrenal gland in stress. The other five had stable or decreased levels of cortisol. The six whose cortisol levels increased over the four years had difficulty with certain memory tasks such as navigating a maze or remembering a list of words. Those whose cortisol levels remained low or even decreased somewhat performed these tests with no difficulties. A subsequent fMRI study showed that the hippocampi of subjects with higher cortisol levels were smaller by about 14 percent.

Retinitis Pigmentosa: A Model Neurodegenerative Disease

As people age, they often lose not only cognitive function but also visual, auditory, and other sensory functions. They can gradually become isolated from other humans and the environment. A particularly devastating condition is age-related macular degeneration (AMD), which robs older people of their central vision. The macula is a specialized region of the retina having the highest density of photoreceptors. At its center is a small indented area, the fovea, which contains only cone photoreceptors and serves all of our high-acuity vision. We look at things we want to examine closely with that small region of the retina. There are only about 35,000 photoreceptors in the fovea, out of a total of 6 million cones in the entire human retina and perhaps 95 million rods, but if the foveal photoreceptors are lost, which happens in AMD, high-acuity vision is lost, and it is devastating for those affected. They cannot read, watch television, or do any of the things normally sighted people take for granted.

We know very little about AMD, what causes it or predisposes to it. The only environmental link known is smoking: Those who smoke have an increased risk of AMD. Some rare forms are inherited and those susceptible might get the disease at a young age. In one such form of macular degeneration, called Malattia Leventinese, the mutation has been identified in a gene coding for a secreted protein of unknown function. The mutation results in misfolding of the protein that causes it to be abnormally secreted and to accumulate both intracellularly and extracellularly.

This form of macular degeneration, then, has characteristics of other neurodegenerative diseases such as Huntington's, Alzheimer's and Parkinson's diseases in which excess accumulation of protein, either extracellularly or intracellularly, is the defining feature. However, for most cases of AMD, the link with genetics is tenuous. At present, there are no animal models of the disease, because relatively few animals have a fovea like ours. Higher primates, some birds, reptiles, and fish do, although the foveas of these latter animals are somewhat different. Thus, we can do relatively little for most cases of AMD at present and it severely compromises the quality of life of those who suffer from it. The same is true of those who become deaf in old age. They, too, can become quite isolated from their fellow humans.

An inherited retinal degeneration called retinitis pigmentosa (RP) is not a disease of the aged. Indeed, in most cases its onset is in the late teens or early twenties, progressing to complete blindness in the fifties, sixties, or even later. It is a fairly rare disease, affecting about one in 4,000 people worldwide. It is a disease of the rod photoreceptors, but cone photoreceptors are eventually affected as well. In contrast to AMD, where central vision is lost, RP begins in the periphery of the retina, gradually restricting the visual field. The fovea is the last to go, but eventually it, too, degenerates, leaving the individual completely blind. This disease has been studied intensively for more than 40 years and enormous progress has been made in understanding its causes. Furthermore, progress is being made with therapies for the disease, at least in treating animal models of the disease, of which there are many. Animals that have RP-like diseases include mice, rats, dogs, and cats.

RP, like many neurodegenerative diseases, was thought at one time to be a single disease with a single cause. We now realize that this is a very incorrect view. The first hint that RP represents several different diseases came from genetic studies. Careful analysis of patients and their families with RP first showed that about 50 percent of the RP cases could be linked to a genetic cause, but the genetics is varied. Sometimes the disease is inherited as an autosomal dominant disease—this means that the chances of an offspring having the disease are 50 percent; sometimes it is inherited as a recessive disease—if both parents have the mutant gene, one out of four children will inherit the gene.

There are also cases where the disease's inheritance is sex linked. The mutant gene is on the X chromosome, but females have two X chromosomes while males have just one. Males show the disease when the gene in their single X chromosome is defective; females need to have the defective gene in both of their X chromosomes. Therefore, males inherit the disease much more frequently than do females in these families. There are some even rarer inheritance patterns in RP, but they need not concern us here. Suffice it to say that the genetic variability is large, but there is much more to come.

Of the 50 percent of RP cases that can be linked to genetics, about 40 percent show an autosomal dominance inheritance. These cases were the first in which the specific genetic defect was discovered in the early 1990s by Thaddeus Dryja and Eliot Berson and their colleagues at the Massachusetts Eye and Ear Infirmary in Boston. They showed that a mutation affecting a single amino acid in the gene coding for rhodopsin (the protein that, when combined with a vitamin A derivative, is light sensitive and initiates vision) was the culprit in one family with RP. This discovery opened the floodgates and within a decade numerous mutations both in the rhodopsin gene and other genes found in photoreceptor cells and important for initiating vision were discovered.

As of this writing (Fall, 2003) more than 100 different mutations in 35 different RP-causing genes have been found. In rhodopsin alone, more than 70 RP-causing mutations have been identified. Thus, RP is not a single disease, but more than 100 diseases, and this accounts for only about 60 percent of the cases of RP. The lesson here is that it is likely many of the neuro-degenerative diseases will be as heterogeneous, and careful classification of them is critical if we are to deal with them. RP might be leading the way in this regard.

Figure 6-4 shows the rhodopsin protein, consisting of a chain of 348 amino acids, each indicated by a circle. The chain weaves in and out of the membrane seven times. The amino acids that, when altered, lead to RP are indicated in black.

At least 70 mutations in the rhodopsin molecule can lead to RP diseases. Some of the mutations lead to a misfolding of the protein, others to alterations in how the molecule is excited by light. Different mutations result in somewhat different diseases in terms of age of onset and progression of the disease, although there is considerable variability in people with the same mutation. This variability or penetrance is not well understood; as discussed in earlier chapters, both genetic and environmental factors could be involved.

With some understanding of the genetic basis of the autosomal dominant form of RP, investigators have turned to therapies and are focusing on two approaches: genetic and pharmaceutical.

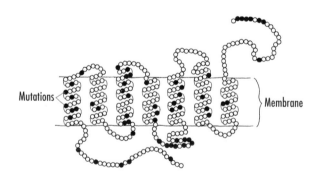

FIGURE 6-4 *A schematic drawing of the rhodopsin molecule. Amino acids which when altered as a result of a genetic mutation cause retinitis pigmentosa are indicated by black circles.*

Gene therapy has already had spectacular success in a strain of dogs with one rare form of RP. These dogs have a defect in a gene that codes for a protein required to make the correct form of vitamin A to make rhodopsin light sensitive. A good gene is inserted into a virus that is then injected into the eye—into the subretinal space between the photoreceptors and the cells behind, the pigment epithelial cells. In this case, the defective protein is present in the pigment epithelial cells, but this is unimportant for the discussion; the defective protein could as well be in the photoreceptors. The injected virus infects the pigment epithelial cells and the cells begin to make the normal protein—they now have the correct gene—and the dogs show a remarkable recovery. For any of this to happen requires, of course, exact knowledge of the mutated gene, a virus that readily infects the defective cells, and luck!

The second approach is via drug or chemical therapy, and several classes of compounds have been tried. Significant success has been obtained in animals with some forms of RP following the administration of growth factors such as the neurotrophins described in the beginning of Chapter 2. These molecules are important for the survival and growth of neurons in the developing brain, and they also seem to help defective cells from dying or, at the very least, they slow down the deterioration of genetically defective cells. All forms of RP might not respond to growth factors, and infusing the growth factors into the eye is a problem because these are proteins that must be put directly into the eye. That we have animal models with various forms of RP is important to this effort, and it is now possible to make mice with other forms of RP using transgenic techniques, thus providing the opportunity to test drugs on a wide variety of mutations that cause the disease.

Another substance that has been shown effective in RP is vitamin A. Initial studies with a population of human RP patients who had various forms of the disease suggested a very small but positive effect of vitamin A therapy. However, the effect was so minimal that many physicians felt it was not particularly useful. But as transgenic mouse models of various forms of RP have been developed and the specific genetic defects found in human patients are induced in them, it turns out that some types of RP are helped quite significantly by vitamin A, but others not at all.

The lesson here is that when seeking therapies for a disease, it is critical to know what specific form of the disease you are dealing with. A therapy might be very effective for one or a few forms of a disease, but ineffective for many or most forms. Thus, when testing a therapy on a population with various forms of a disease, the positive effects might be swamped out by the nonresponders. Thus, careful characterization of different forms of the neurodegenerative diseases is essential, and this classification is just beginning for many age-related neurodegenerative diseases such as AMD, Alzheimer's, and even perhaps Parkinson's disease.

CONCLUSIONS
(AND SPECULATIONS)

Neurobiological studies of the developing brain provide much information on how the brain initially forms in the fetus. At first glance, we might conclude that early brain development depends strictly on nature—intrinsic genetic directives—and Chapter 1 appears to support this view. But it is important to recognize that environment and nurture can also play a role in early brain development. I use the term "environment" here and in the rest of this discussion on brain development very broadly. Essentially, I mean nongenetic factors, of which environment is only one, although perhaps the major one. As I discussed in Chapter 5, random developmental variations due simply to chance could and probably do occur during development, and affect brain development to some extent. In most instances, this might make little or no difference, but in others it could well make a substantial one. We simply don't yet know. However, certain environmental factors that perturb early brain development are easy to document, and some can have devastating effects.

An obvious and dramatic example of the substantial role that environment can play in early brain development is that of fetal alcohol syndrome (FAS). Children born to alcoholic mothers show a wide range of developmental brain disorders, from misaligned cortical cells and abnormal clusters of cells in parts of the brain to an absence of many of the cortical infoldings and a significantly undersized brain. Severely affected children are dramatically retarded mentally, and less affected children demonstrate learning disabilities, lower IQ scores, and behavioral problems, including hyperactivity.

How much alcohol consumption is required to cause such problems? No one knows for sure, but binge drinking, especially early in pregnancy, seems to result in the most severe cases of FAS. And it is easy to show that just a few drops of vodka added to their surrounding water cause zebrafish embryos to develop significant brain malformations.

Coupled with alcohol consumption in causing severe FAS is the nutritional state of the mother and the use of other drugs, including tobacco. Thus, early brain development can be influenced by a variety of environmental factors, including the mother's health, her diet, and perhaps even her level of anxiety and or stress. In support of this notion is the evidence that socioeconomic status is the best predictor of health, longevity, and absence of mental illness in all societies. This is not a very well studied area, but it needs to be kept in mind when thinking about the relative roles of nature and nurture in early brain development.

Environmental influences on a fetus might be subtler than the examples given above. One accepted notion put forward to explain the differences between identical twins is that the *in utero* environment can be different for different fetuses. Some fetuses might receive slightly more or less nutrition because of a somewhat different blood supply to the fetus or perhaps where a fetus resides *in utero* at a particular time could make a difference. The observation that infants at birth prefer the language spoken by their mothers (discussed in Chapter 3) suggests that even sensory input *in utero* might influence brain development to some extent. The question is how much do these subtle factors matter? We simply don't know the answer. And of course, every pregnant woman wants to know what she should do to optimize her baby's health and future happiness, but again we can't say what. We can describe things not to do, but this is as far as neurobiological facts can take us.

We must grant, though, that in healthy mothers, nature—intrinsic genetic directives—is of primary importance in establishing the framework of the developing brain, though framework is probably not the best word or correct concept to describe early brain development. Indeed, the evidence is that the brain substantially develops—even overdevelops—by these intrinsic genetic directives. Sophisticated neuronal circuits are formed by intrinsic mechanisms, and remarkably adult-like responses can be elicited from neurons in newborn, environmentally inexperienced, brains. The visual system results described in Chapter 2 make this point well.

This does not mean that intrinsic directives wire everything up precisely. Refinement of circuits clearly involves experience, and early in its development the brain is particularly amenable to modification, modulation, refinement, or whatever you might wish to call it. These early times of exceptional plasticity are the critical and sensitive periods.

What we know about maturation of the brain (Chapter 2) might surprise some and is perhaps an area where educators and others might be influenced by the neurobiological evidence. The sculpting of the brain during its maturation phase consists to a considerable extent of a pruning and refinement process. The young brain has more neurons, more expansive branching patterns, and more synapses than the adult brain, and environment—nature—plays a critical role in the refinement and pruning. In birds, for example, at the end of the critical period for either vocal learning or imprinting the density of synaptic spines on the key neurons in the appropriate nuclei drops to about half of what it was during the critical period. (Chapter 3 discusses these changes.)

Thus, the amount of potential synaptic input into these neurons is significantly reduced after the critical period is over.

The unexpected conclusion is that the brain initially has great intrinsic capability and potential, and during brain maturation some capability is lost—the adage "Use it or lose it" fits here. The game, then, is to work toward losing as little brain capability as possible. Language acquisition is the model here (discussed in Chapter 3). Young infants can distinguish and make the sounds of any language, but they lose this ability within the first few years. Children readily learn languages early on, but by puberty it becomes more difficult for virtually everyone to learn a new language. Should we be exposing our youngsters to the sounds of many different languages early on, and should we begin language instruction much earlier than we presently do? We don't know the answers, but they seem worth considering.

This general principle for language acquisition holds for other capabilities as well, from learning to throw a ball to playing a musical instrument or manipulating a computer. Encouraging youngsters to develop skills early would appear to make sense from what we now know. The example here, of course, is the observation that string players who learned to play their instrument before the age of 12 have a greater cortical representation of the left fingering hand than do musicians who began to play later in life. The point is that the young brain is more plastic, more modifiable than the adult brain, and perhaps we should take advantage of this property.

But a critically important question is how far can we push the envelope? How far can experience go in taking advantage of early brain capability or, to go further, can the brain's capability be expanded beyond what is there initially? The experiments with rats and enriched environments indicated that it is possible to induce the sprouting of new processes and the formation of new synapses in the young animal, but this occurs, fortunately, over one's entire lifetime and is not limited to the young brain, as discussed in Chapter 2. By raising animals in enriched environments, new brain circuitry can be induced to form, but it is superimposed on a massive pruning and refinement of neural circuitry that is naturally occurring. The owl experiments described in Chapter 3 suggest that new synapses and circuits formed early in an animal's life—during the critical or sensitive period—might remain into adulthood even if they are not used for a considerable time and even if they have become entirely silent.

What the neurobiology is telling us—the bottom line—is that genetic directives are clearly most critical in brain building, although the environment can also play some role, whereas environmental factors play the fundamental role during brain maturation, although there is genetic restraint. This does not mean that environmental factors during brain maturation can greatly override the brain's intrinsic capability. We all differ significantly because we are different genetically. The view of behaviorist John Watson in the 1920s that he could turn any healthy infant into a "doctor, lawyer, artist, merchant-chief and yes, even beggar-man and thief" by environmental influences is not accepted by any serious scientist today. Each of us has different capabilities and talents and this certainly reflects to a great extent our genetic makeup. But within that genetic makeup, there is room for modification, even perhaps for some elaboration, and this is where experience and environment come in. Of course, these are extraordinarily contentious issues, not because most people today do not agree that what we are is a mix of nature and nurture, but because we are not sure how much each contributes to the final product. This is where the great sticking points lie, although attempts to put numbers on the extent that behavior or capability is genetically or environmentally based are continually being made. (See Chapter 5 for a discussion of this.)

Neurobiology contributes little to this debate, except to say that both nature and nurture are clearly involved. But to reiterate, what we have learned neurobiologically about brain development should guide us as we raise and educate our children.

Genes and Behavior

Unequivocal examples of individual genes causing specific neurological diseases that significantly alter behaviors are now known, Huntington's disease being one (discussed in Chapter 5). A dominantly inherited disorder, it occurs in everyone who inherits a sufficiently defective copy of the gene. The nature of the gene defect in Huntington's disease is now understood, and the previously unexplained variation in onset and progression of the disease observed in those suffering from it appears to relate mainly to the extent of the defect in the gene. That is, the defective gene has an excessive number of CAG repeats, and the more repeats, the earlier the onset and the faster the progression of the disease.

That individual genes can exert different phenotypic effects on individual organisms has long been appreciated and is usually termed gene penetrance. It is often ascribed to environmental or epigenetic effects on gene expression, and this might be true in many cases. However, in the case of Huntington's disease, gene penetrance is explained to a considerable extent by variations in the defective gene itself. It might also be explained by variations in normal genes in an individual—so-called polymorphisms. These are alterations in genes that produce proteins that function quite normally but that alter the response of a tissue or organism to a particular environmental condition. Let me illustrate with a dramatic example. Rodents, especially albino ones, are quite susceptible to light damage of their photoreceptor cells. If continuously exposed to ordinary room lights for just a few days, the animal's photoreceptor cells degenerate. A surprise observation made a few years ago was that one strain of albino mice is highly resistant to light damage. Much more continuous light exposure is required to cause photoreceptor damage in these animals compared to most strains of mice. Comparing the photoreceptor responses of this strain to others reveals no very significant differences; they all seem to function within normal limits. The variation shows up only under the stress of continuous light.

A genetic difference between this and other strains has now

been uncovered. It is the result of a single amino acid change in a protein needed to make the correct form of the vitamin A derivative bound in rhodopsin (discussed in Chapter 6). In the resistant strain, the correct form of the vitamin A derivative needed to make rhodopsin is not made quite as fast; thus, after being broken down by light, a normal event, rhodopsin is not reformed as quickly in the light-resistant strain as in light-sensitive strains.

This change probably has little effect on the visual performance of the animals. Indeed, the resistant animals make as much rhodopsin as do the light-sensitive ones and their photoreceptors can detect dim light stimuli as efficiently as those in other mouse strains; it just takes the light-damage-resistant mice somewhat longer to reach this level of performance. It is only under conditions of continuous light that the retinas of the light-resistant and light-sensitive strains respond very differently—and because of a tiny—one amino acid—difference in one protein.

The point here is that very different phenotypes under specific environmental conditions can result from what might be considered insignificant genetic differences. The relationships, then, between genes, their products, and the environment are complex and not easy to sort out.

That a number of neurological diseases such as Alzheimer's disease and cognitive diseases such as schizophrenia have links to genetics is not at all surprising. Indeed, it might be inevitable, but the nature of the genetic link is the critical question. We talk of predisposing genes for such diseases, but exactly what that means in many cases is difficult to define. In the case of Alzheimer's disease, the predisposing genes all appear to be related to the synthesis or breakdown of β-amblyoid, the protein that accumulates in the brains of sufferers and is its precipitating cause (discussed in Chapter 6). This makes sense, and if we propose that there might be genes that predispose someone not to be susceptible to a disease, we might then be able to explain the Aunt Marians who live to be 102 and remain perfectly normal cognitively. The example described earlier, of a polymorphism in a protein that makes photoreceptors resistant to breakdown in continuous light, could be viewed as the product of a predisposing gene that acts like that—to counter an environmental stressor and prevent neuronal degeneration.

At least an order of magnitude more difficult to answer is the question of genes and cognitive behaviors. As noted in Chapter 5, whereas claims have been made for individual genes controlling, or even strongly predisposing people to a specific complex behavior, none of these claims have held up in a convincing way. It is almost certainly true that there are predisposing genes for cognitive behaviors, but this has not yet been pinned down, and for any such behavior there are, in virtually all cases, multiple genes involved—pulling and pushing in opposing directions. It is no wonder, then, that the field of behavioral genetics is in a muddle as far as complex cognitive behaviors are concerned. Some believe that we will never be able to relate complex behaviors to genetics in any meaningful way because of the complexity and obviously large role that environment must play.

A recent article in *Science* magazine entitled "Rethinking Behavior Genetics" by Dean Hamer, a behavioral geneticist at the National Institutes of Health, reflects the frustration of those in the field. He ends his article with the following:

> Human behaviors and the brain circuits that produce them are undoubtedly the product of intricate networks involving hundreds to thousands of genes working in concert with multiple developmental and environmental events. Further advances in the field will require the development of techniques, such as microarray analysis, that measure the activity of many different genes simultaneously. Only then will the gene hunters have a shot at achieving the promise held out by the past century of classical behavior genetics research.

But it is perhaps useful to point out some of the remarkable similarities in identical twins raised apart and studied by Thomas Bouchard before completely dismissing the idea that the study of human behavioral genetics is irrelevant. One of the first pairs of identical twins studied by Bouchard were boys separated five weeks after birth and raised in different families about 80 miles apart in Ohio. When they were reunited after 39 years, the similarities between them were remarkable. They both were 6 feet tall and both weighed 180 pounds, but more surprising was the striking similarity of many of their behavioral characteristics. They had the same walk and many identical mannerisms—from the way each picked up a knife to nail-biting. They had similar likes and dislikes—from stock-car racing (like) to baseball (dislike). Their houses were similar in design and size and each had an elaborate workshop where he made wooden objects similar to those made by his twin. As far as these two were concerned, it was harder to find differences than similarities in their behavior and personalities.

Because these twins were raised in the same state, less than 100 miles apart, it might be supposed that proximity could account for at least some of the remarkable similarity between them. But another set of male twins, split apart only a few months after birth, were brought up in very different environments—one in Trinidad and the other in Germany. They first met at age 21, but then had very little communication until they were reunited in Minneapolis in the early 1980s when they were about 50 years old and were studied by Bouchard and his colleagues. Again, some of the similarities between these two were astonishing. Their gaits were similar; they had unusual habits in common such as storing rubber bands on their wrists and reading magazines from back to front. There were certainly differences between them, but the similarities in mannerisms and temperament were striking. Sets of identical female twins raised apart showed similar mannerism identities, from excessive giggling to one set of twins arriving in Minneapolis with each having seven rings on her fingers.

What are we to make of these curious similarities? No one is sure, and other investigators have described identical twin pairs raised in homes differing in social class as having quite different behavioral traits, but I don't think the above examples can be easily dismissed as chance. They would seem to be genetically based, but how? One would imagine that such trivial personality traits would reflect environment much more than genetics and, if genetics, an exceptionally complex genetics that would not likely result in such obvious similarities.

What Does the Future Hold?

A major realization of the past two decades is that the adult brain is more modifiable than previously believed (discussed in Chapter 4). That we can learn and remember things our entire lives has long been recognized, of course, but this was viewed as the exception, not the rule, as far as modifiability of the adult brain is concerned. Today the view has softened—not that we believe the adult brain is as plastic as the young developing brain, but we do think it is possible for the adult brain to acquire abilities previously thought unavailable to it.

This new realization has encouraged researchers to seek ways to allow the adult brain to achieve skills ordinarily managed only by the developing brain. One undertaken by Jay McClelland and his colleagues at Carnegie-Mellon University is to teach Japanese adults to distinguish "r" from "l" sounds which they have difficulty doing (see Chapter 3). McClelland and colleagues have reported some success, albeit with only a few subjects. They did this by first presenting to the subjects exaggerated and even distorted speech sounds that never occur normally. As the subjects began to discriminate these sounds, they were gradually presented with more normal, harder to discriminate sounds. Whereas initially the subjects could discriminate the sounds at levels only just above chance (that is, 50-60 percent), after 480 training trials, the subjects improved to 80-100 percent correct discriminations. Obviously this preliminary study needs to be expanded and repeated, but it is promising, and other, more effective, ways might be found to achieve such results.

Another approach being undertaken is to study that small cohort, less than 5 percent of the population, that learns second languages very effectively as adults. What is different about these people's brains, and how do they go about learning a new language? Can any light be shed on the issue by studying them? As yet no definitive answers are available.

A third approach is to carry out such studies in animals, and a recent report by Knudsen and his colleagues at Stanford suggests that it is possible to achieve some compensation in adult owls when prisms that shift their visual field are placed on the animals, something thought not possible after the critical period for this plasticity had passed. (Knudsen's work with owls is discussed in Chapter 3.) The key here was to shift the visual field by only a small amount at a time. Using such a training paradigm, the adult owls showed some compensation. The extent of compensation was limited compared to young owls, but that some plasticity could be induced was unequivocal and interesting.

Some ocular dominance plasticity has now been observed in the visual cortex of adult mice also. Mice, unlike cats, monkeys, and ourselves, have only a small area of visual field overlap in the two eyes because their eyes are on the sides of their head and do not point forward. In the area of visual field overlap, inputs from the opposite-side (contralateral) eye to the cortical neurons predominate, although weak input from the same-side (ipsilateral) eye can be detected. By occluding the dominant eye by lid suture and extending the period of deprivation, strengthening of the ipsilateral input to the cortical neurons was found. Interestingly, this cortical plasticity depended on the presence of NMDA receptors; the ocular dominance plasticity was not observed in mice that had the NMDA receptors knocked out genetically. As discussed in Chapter 4, these glutamate receptors are critical for the generation of long-term potentiation not only in memory and learning but in other forms of cortical plasticity as well.

How far we can go in training the adult brain is, of course, not at all clear, but the new data are certainly encouraging and recommend that we rethink the issue. Approaches might involve not only training normal adult brains but also retraining damaged brains. Are we too quick to decide that nothing can be done following a stroke or other serious neurological conditions? I noted in Chapter 5 the devastating injury to the actor Christopher Reeve, whose spinal cord was crushed in a riding accident. Whereas it was generally believed that his injury was permanent and nothing could be done to help him, some novel treatment approaches applied to him appear to have resulted in surprising progress. The reports so far have appeared mainly in the media but, if confirmed, suggest that we might be able to do much more than previously thought for such serious neurological injuries.

At the same time we are beginning to achieve some understanding of the neurobiological factors involved in promoting neuronal cell survival or inhibiting neuronal cell death as well as promoting axonal regeneration. As this work progresses, it is likely that new therapies will become available to deal with neurological injuries and disease.

In Chapter 6, I argued that a biological limit to maximum human life expectancy is likely and that within a few years average life expectancy will reach a plateau, at least in the developed countries. The reason, according to biodemographers, that average life expectancy will plateau is that many of the causes of early death—especially infectious diseases—have been dealt with. Furthermore, there has been substantial progress in reducing early death from the other major killers, including cardiovascular disease, diabetes, and cancer.

My own view is that our life span is determined mainly by our brain. That neurons are not replaced in the brain for the most part and that brain structure and function gradually deteriorate with age seem unequivocal and the ultimate determinant of a finite life span. As noted in Chapter 6, it is possible to transplant hearts, livers, and kidneys as well as other organs from humans and even animals, and artificial organs are being developed. But I don't think anyone seriously believes that we can transplant a whole brain or make an artificial brain. Indeed, even if one could do this, the uniqueness of that individual would be destroyed. Furthermore, as noted earlier, if whole brain transplantation were possible, it would be better to be the donor than the recipient!

It is conceivable that we will find ways to replace neurons with stem cells, either those that exist in certain brain regions or others that are transplanted into the brain, but I think these possibilities are still remote and, even if they do become feasible, would they ever be able to maintain or replace an entire brain? And, of course, is this something we would even want to do—to prolong human life to 150-200 years or longer? (Chapter 6 discusses this.)

I am not suggesting that we should stop trying to cure neurodegenerative diseases or to find ways to replace dead or dying neurons with stem cells. But our goal in these studies should be to improve the quality of life for those in their later years, not to increase maximal life span. One might relate to the other, but not necessarily so, and it is the former goal—to optimize the years we have to spend on this planet—we should strive for.

To end this book on a more positive note, let me emphasize again that neuroscience as a field has progressed spectacularly over the past half century. Much of this progress has been at the cell and molecular levels. We now have quite a good grasp of how individual neurons function—how they receive, integrate, and carry signals and how they pass on information to other cells. The field is now turning to a systems-level analysis—how aggregates of neurons interact to underlie behaviors. These studies provide the links with psychology and promise to give us an understanding of the brain, behavior, and a number of the issues described in this book.

In this quest, it is still early days, and it might still be asking too much of neuroscience to provide definitive answers to such contentious issues as the nature-nurture debate in brain development or the relative roles of genetics and environment in human behavior. I have emphasized the point over and over that neuroscience at the moment can take us only so far. However, I think that neuroscience has given us some glimpse of how many of these questions might be answered and even, perhaps, models to ponder.

Further, the future for much more progress is bright. Several noninvasive techniques for studying the human brain—PET scanning, fMRI, and magnetoencephalography and their variants—are available. And we can already analyze what is going on in animal brains down to the single synapse. Combining the two approaches is powerful and is key to providing a compelling picture of how the brain works, how best to encourage its development, and how best to maintain it.

FURTHER READING

These recent books that touch on various topics discussed in this book are accessible to nonscientist readers and most are a good read.

Blum, Deborah. *Love at Goon Park: Harry Harlow and the Science of Affection.* Cambridge, Mass.: Perseus Publishing, 2002.

Bruer, John T. *The Myth of the First Three Years: An Understanding of Early Brain Development and Lifelong Learning.* New York: The Free Press, 1999.

Dowling, John E. *Creating Mind: How the Brain Works.* New York: W.W. Norton & Company, 1998.

Eliot, Lise. *What's Going On in There?: How the Brain and Mind Develop in the First Five Years of Life.* New York: Bantam Books, 1999.

Gopnik, A., Andres N. Meltzoff, and Patricia K. Kuhl. *The Scientist in the Crib: What Early Learning Tells Us About the Mind.* New York: William Morrow & Company, Inc., 1999.

Kolb, Bryan, and Ian Q. Whishaw. *An Introduction to Brain and Behavior.* New York: Worth Publishers, 2001.

McEwen, Bruce (with Elizabeth Norton Lasley). *The End of Stress as We Know It.* Washington, D.C.: Joseph Henry Press, 2002.

Pinker, Steven. *The Blank Slate: The Modern Denial of Human Nature.* New York: Viking Penguin, 2002.

Restak, Richard. *The Secret Life of the Brain.* Washington, D.C.: The Dana Press and Joseph Henry Press, 2001.

Ridley, Matt. *Nature Via Nurture: Genes, Experience, and What Makes Us Human.* New York: HarperCollins Publishers, Inc., 2003.

Sanes, Dan H., Thomas A. Reh, and William A. Harris. *Development of the Nervous System.* San Diego: Academic Press, 2000.

Wright, Lawrence. *Twins: And What They Tell Us About Who We Are.* New York: John Wiley & Sons, Inc., 1997.

FIGURE CREDITS

1-2 After Cowan, W. M., *Sci. Am.*, **241**, 106, 1979.

1-3 From Purves, D. (1994), *Neural Activity and the Growth of the Brain*, Cambridge University Press, Cambridge, England.

1-4 From Purves, D. (1994), *Neural Activity and the Growth of the Brain*, Cambridge University Press, Cambridge, England.

1-6 After Rakic, P., *Science*, **183**, 425, 1974.

2-1 After Purves, D., and Lichtman, J. W. (1985), *Principles of Neural Development*, Sinauer Associates, Inc., Sunderland, MA.

2-2 After Snider, W. D., *J. Neurosci.*, **8**, 2628, 1988.

3-2 After Konishi, M., *Annu. Rev. Neurosci.*, **8**, 125, 1985.

3-3 After Knudsen, E. I. (1999), in *Fundamental Neuroscience*, M. Zigmond *et al.*, eds., Academic Press, San Diego, p. 637.

3-4 After Knudsen, E. I. (1999), in *Fundamental Neuroscience*, M. Zigmond *et al.*, eds., Academic Press, San Diego, p. 637.

4-2 After Merzenich, M., and Jenkins, W. M., *J. Hand Therapy*, **6**, 89, 1993.

4-3 After Ramachandran, V. S., *Proc. Natl. Acad. Sci.*, **90**, 10413, 1993.

4-10 After Wurtz, R. H., *et al.*, Sci. Am., **246**, 124, 1982.

5-2　After Restak, R. (2001), *The Secret Life of the Brain*, Dana Press (New York) and Joseph Henry Press (Washington, D.C.).

6-1　After Strehler, B., *et al.*, *J. Theor. Biol.*, **33**, 429, 1971.

INDEX